Finding Your Way Through Formal Verification

@2023 by Bernard Murphy, Manish Pandey, Rajeev Ranjan, and Sean Safarpour

All rights reserved. No part of this work covered by the copyright herein may be reproduced, transmitted, stored, or used in any form or by any means graphic, electronic, or mechanical, including but not limited to photocopying, recording, scanning, taping, digitizing, web distribution, information networks, or information storage and retrieval systems, except as permitted under Section 107 or 108 of the 1976 US Copyright Act, without the prior written permission of the publisher.

Published by SemiWiki LLC Danville, CA

Although the authors and publisher have made every effort to ensure the accuracy and completeness of information contained in this book, we assume no responsibility for errors, inaccuracies, omissions, or any inconsistency herein.

First printing: March 2018 Printed in the United States of America

Second printing: January 2023 Printed in the United States of America

Table of Contents

Table of Contents 2
Foreword to the Second Edition 5
Introduction to the Second Edition 8
Chapter 1 Time for a Fresh Look at Formal 9
    Why now? 9
    Why another book on this topic? 11
    Organization of this book 13
Chapter 2 The Verification Treadmill 16
    Why verification is so important 16
    The price of not being perfect 18
    Why is this so hard? 19
    Battling the exponential 22
    Enter formal verification 27
Chapter 3 Formal Verification – the Early Years 29
    Background 29
    Equivalence checking 33
    What do we mean by correctness? 35
    Properties: assertions and assumptions 39
    Under the hood: the formal engines 43
Chapter 4 What's the Catch? 50
    Property checking goes commercial 50
    A problem with assertions 50
    Completeness is expensive 51
    Inconclusive and debug 53
    Stuck at the chasm 55

Chapter 5 Crossing the Chasm – Fearless Formal 57
Pre-packaged solutions 57
Auto checks – functional linting 59
X-propagation verification 61
Coverage analyzer – formal helping simulation 62
Connectivity checking 63
Clock domain crossing 67
Functional safety 68

Chapter 6 Apps Requiring More Hands-On Experience 70
Sequential equivalence 70
Register verification 73
Low power 75
Security verification 76
Testbench analyzer 78

Chapter 7 Formal Methods for Datapath Elements 80
The exploding importance of custom datapath functions 80
Verifying datapath elements is a challenge 82
Back to equivalence checking – with a twist 84
Practical considerations in datapath verification 86
Recap 88

Chapter 8 The Role of Formal in Design Today 89
Adoption 89
Size constraints 91
Popular targets for property checking 92
    Good targets 92
    Targets to avoid 95
Product and service solution providers 95

Chapter 9 Adding Formal to Your Flow 97
    Organic skills growth 97
    First-cut targets for formal verification............................... 100
    Detailed planning.. 102
    Monitoring progress 103
        Under-constraining and over-constraining 105
        Bounded proofs (inconclusives) 106
        Manually guided proofs.. 108
    Getting to formal signoff 109
Chapter 10 Looking Forward.. 112
    Extending formal methods to C/C++ specifications 112
    Cloud-based formal verification 113
    Machine learning and AI 113
    Architectural formal verification 114
    Cross-leveraging simulation and formal investments........... 116
    The outlook for formal verification...................................... 116
If You Want to Go Deeper... 118
Glossary of Formal Terms 119
Acknowledgements 123
About the Authors 124

Foreword to the Second Edition

Five years ago, the authors directed the first edition of this book at people who were curious about formal, then a fairly significant subset of the design verification community. These were organizations thinking about formal or experimenting but still on the fringes of commitment. Five years later, what has changed? I think it's fair to say that curiosity has almost everywhere grown into active adoption – to more design organizations and into broader involvement in the total verification objective. As an early adopter of formal methods for semiconductor design it is not surprising that Intel is at the leading edge of this expansion. But talking to friends and colleagues in many other semiconductor and systems companies I see the trend accelerating everywhere.

Some of this expansion can be attributed to improved ease of access and quantification of results. The 'app' concept has dramatically shortened the learning curve to productive verification, even for relative novices. Improved definitions of coverage have increased trust that formal testing can be measured just as effectively as for dynamic testing. In addition, property checking continues to prove an almost unbounded ability to tackle hard problems beyond the reach of simulation. Accumulating successes prompted an obvious question – while formal is obviously a valuable verification technique, is it only for special cases or is it possible to significantly improve the contribution of these methods to the total verification task?

We took that question as a challenge to advance to (block-level) formal signoff. If formal can replace dynamic for signoff at lower cost, that would be a significant contribution. If it can guarantee a higher quality signoff, that would be even more valuable. More aggressive application of commercial formal methods, also specification level comparisons through C to RTL equivalence checking have been very important in meeting this goal. We started this transition several years ago and now have multiple blocks we signoff only in formal. We find for other blocks we can find the majority of bugs and signoff those areas with formal, leaving a still important but diminished role for dynamic debug and signoff.

The historical limit of formal verification to control logic would have limited scope of this idea if it were not for formal datapath verification. Considering the importance of datapath elements in GPU, DSP, AI and many other accelerators today, breaking through the datapath boundary has been a critical step towards drawing a great majority of unit signoff tasks under the formal umbrella. This change, from dynamic to formal signoff, has unquestionably given us a significant productivity advantage. But more than that, we're proud that some of our most important blocks signed off with this method have not exhibited a single bug across multiple product generations. Higher productivity and higher quality – for us the ROI argument for formal signoff is unquestionably proven.

Another area we extend the ROI of formal methods is in architectural verification, something we have recently adopted with significant success. Applications here include proving the correctness of a coherent mesh fabrics, also proving correctness of firmware running on a CPU cluster. Today these techniques depend on a combination of our own development and expertise in abstracting the design, together with open-source tools and some commercial products. I'm looking forward to seeing more of this capability standardized over time.

I found it interesting that the authors talked quite a bit in the first edition about the organizational aspects of formal verification. These have become even more important. In growing a pool of experts, it is still uncommon to see fresh EE graduates with useful understanding of formal methods. Yet the demand from industry is now intense. Many large companies I know have teams of 100+

engineers dedicated purely to formal. Even the most advanced small companies are stretching to add more formal capacity.

This demand can only accelerate as we move to more formal signoff, more architectural formal and more hardware/software formal validation. Core verification training in dynamic methods in most EE programs is not a good preparation for the formal mindset these graduates will need. Across the industry we are having to invest in basic training to help new grads adjust. On a positive note, I equally find that once trained, formal engineers are even more passionate about what they do than others. The nature of formal verification demands that they understand a design in more detail than might be required for dynamic testing. This is a perfect recipe for engineer retention and engineer personal growth!

The outlook for formal verification could not be brighter. Simulation still has a very important role to play but is already on its way to being displaced for unit level signoff. It is not unreasonable to guess that five years from now, simulation will be used only for subsystem and system level verification. Meanwhile, formal methods have started to tackle some system level tasks. Today these methods depend heavily on in-house support; five years from now I expect more of these methodologies covered in standardized commercial support.

Organizationally, formal verification engineers have the best of all possible worlds – an intellectually challenging domain and a market where they are in high demand. I look forward to seeing more EE/CS undergraduate courses in support of this demand and a rapidly growing cadre of experts who see the challenge and the opportunity for their own careers.

M. V. Achutha Kiran Kumar
Intel Fellow; General Manager, Formal Verification Central Tech Office
2022

Introduction to the Second Edition

In the five years since we released the first edition of this book technology has advanced as it always does. We felt it was appropriate to provide an update, for new readers and for those of you who want to skip to the changed sections. We've made a number of small changes throughout, but the major changes are in the following sections.

Chapter 5: Crossing the chasm – fearless formal. We re-ordered apps to better reflect ease of adoption.

Chapter 6: Apps requiring more hands-on. Many of these apps were introduced in the first edition in the "Looking forward" chapter. Now they are proven and mainstream, they are given more weight in this chapter.

Chapter 7: Formal methods for datapath verification. This topic was also introduced as a somewhat forward-looking capability in the first edition. Since this is a major advance in formal methods, we felt it merited its own chapter.

Chapter 10: Looking forward. What looked new five years ago has inevitably changed. Here we present our updated view of near to medium term possibilities.

Chapter 1 Time for a Fresh Look at Formal[1]

Why now?

You might imagine that the people who build the advanced hardware technologies you find almost everywhere today would feel comfortable with almost any aspect of technology related to their domain. Or at minimum they wouldn't feel intimidated by any topic. They might not understand it now but, if the need for understanding arises, you expect they would be confident that they can quickly become sufficiently expert, as they have already demonstrated through their mastery of multiple verification techniques: static, directed, and constrained-random simulation, along with emulation and prototyping. For rare problems where methods and tools were available but difficult to use, they could always hand a problem "over the wall" to dedicated experts.

It might surprise you to learn that many otherwise expert designers and design managers, if pressed to answer honestly, will admit that they put formal verification in that over-the-wall category and often find it confusing or intimidating. The problem is not so much in broad concepts but in going any deeper, or in knowing how to quantify value. Until relatively recently this wasn't much of a problem. For many, formal verification was at the periphery of the verification toolset. Where a few especially challenging problems defeated conventional verification approaches, they were passed over that wall to experts in formal methods, who would translate reasonable English-language requirements ("we need to check if this can ever happen") to formal tool inputs, then coax the tools into performing their magic and finally provide back either a

---

[1] A quick word about nomenclature: we'll use *formal*, *formal methods* and *formal verification* fairly interchangeably in this book. While this usage is a little loose, it does follow common practice among non-specialists

thumbs-up ("that problem can't happen") or an example of a realistic possible failure.

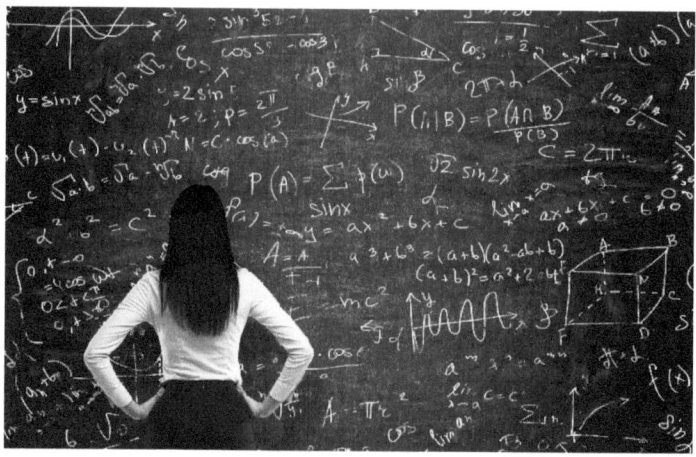

Valuable though this service was, the impact of formal verification in those early days was limited. Even point problems are important to find, but it was difficult to quantify how this technology contributed to overall verification signoff. Formal methods lacked obvious, much less signoff-quality metrics so signoff (is this design production-ready?) clearly remained the responsibility of traditional verification. If there was interest in using formal methods, executives had to consider the cost of building and maintaining a team of specialists, a worthwhile investment for large enterprises (as we'll see) but beyond the reach of more modest budgets.

How times have changed. Now formal verification stands shoulder to shoulder with simulation, static methods, emulation and prototyping, a co-equal in verification flows across all large and many small design and verification organizations. This is partly thanks to continuing improvements in the capability and usability of tools, but more significantly it has been driven by the relentless increase in complexity of modern designs. Some verification tasks, once solved by throwing more bodies, more licenses, more machines at the problem, have already moved beyond the reach of confident signoff through non-formal methods.

Executives are always concerned about the impact of quality problems escaping to the field; they worry especially about critical components exhibiting intermittent problems from product to product. Could one of these latent problems suddenly explode into a customer crisis? Those same executives are now doubly sensitized to the media and market fallout that can result from a publicly exposed safety problem or hack. They are now actively sponsoring teams and methods to mitigate these risks. Formal verification has become prominent in those efforts.

Why another book on this topic?

There are already many books on formal verification, from academic to application-centric, and from tutorials for beginners to guides for advanced users. Many are excellent for their intended purpose; we recommend a few at the end of this book. But most start from the assumption that you have already committed to becoming a hands-on expert (or in some cases that you already are an expert). We feel that detailed tutorials are not the easiest place to extract the introductory view many of us are looking for – background, a general idea of how methods work, applications and how formal verification is managed in the overall verification objective.

There are a lot of us who aren't yet at that commitment stage, or who possibly may never want or even need to become hands-on experts. If this describes you, a 300-400-page tutorial may be more than you are ready to attempt; you want something you can read through relatively quickly to get a general understanding of the domain. This book was written for you to dip a toe in formal waters. If you like what you read, you can move on knowing that an investment in serious learning will be worthwhile. If you don't, hopefully you still feel you have gained enough insight to defend, again more knowledgeably, why a deeper understanding of formal methods isn't appropriate to your current objectives.

Who might this describe? You could be:

- *A Design or Verification Engineer*: You've heard about formal verification, maybe read a little on the topic, or sat in on presentations or tutorials, but you're uncertain whether this direction is right for your needs. You're intrigued by the idea but not quite ready to pick up a textbook; you want to ease into it. This book will get you started with a good broad understanding and should set you up to make that textbook less daunting if that's where you want to go next.

- *A Design or Verification Leader or Manager*: If you're planning to directly manage a formal team, you have to start somewhere. Just like the hands-on engineer, you'd probably appreciate a little orientation before you dive all the way in. Even if you're not directly supervising a formal team, if you're a designer or verification lead or manager, you can expect formal experts to come to you asking questions and looking for guidance about your design, or what is covered in other testing. If such an engineer asks you about an acceptable state-space radius to adequately check a property, you probably would like to know what on earth they are talking about, without having to become a formal expert. We can help.

- *A Verification Manager or Director*: Here we're talking about people who plan and supervise formal verification activity as a part of their overall verification responsibilities. In verification planning, you certainly need to know where formal can play a role and where it may not be suitable, what effort and expertise should be planned for in using these techniques (like most verification techniques, these generally aren't push-button) and how you can assess effectiveness and coverage in what formal teams report back to you. We aim to help with some insights on getting to signoff with formal verification.

There are others we hope will also find value in this book – those of you who are only peripherally involved in verification or who maybe aren't even in engineering. You might be in applications support in a different domain, in sales or marketing, you might be an executive or even in finance or legal. Perhaps you will never run a formal tool or sit in on a verification meeting, but you feel you could be more effective in your role with a better understanding of this domain. As much as others more directly involved, you deserve (if you have the interest) to better understand formal verification and to see where it fits in enhancing product quality. To serve the needs of this broad audience and in the spirit of an introductory overview, we have kept technical detail to a minimum.

Organization of this book

Since we're writing for a fairly wide audience, we cover some topics that some of you may consider elementary (why verification is hard), some we hope will be of general interest (elementary understanding of the technology) and others that may not immediately interest some readers (setting up a formal verification team). What we intentionally do not cover at all is how to become a hands-on expert.

Chapter 2 presents an overview of the verification problem in SoC design, why this is hard and various techniques common in managing complexity, as an introduction to the role that formal methods can play in the larger verification task.

Chapter 3 reviews the early history of formal verification in our industry, along with some of the basic concepts like assertions and constraints and a very lightweight introduction to the engines that drive formal tools.

Chapter 4 walks through the early challenges formal methods faced in getting to widespread adoption: complexities in setup, running and debug and the level of expertise required to effectively use the tools.

Chapter 5 talks about how formal tool suppliers overcame adoption challenges by introducing apps which provide much simpler use-models for targeted applications. This chapter also reviews a number of the most common apps.

Chapter 6 extends the review of apps to several now very popular examples, including clock domain analysis, low power analysis and safety verification.

Chapter 7 expands on a topic covered only briefly in the first edition: datapath verification. This technology is now mainstream and deserves more detailed coverage.

Chapter 8 covers the ways formal is being used today and how wide that usage is. If you want to persuade your manager that an investment in formal is worthwhile, you may find useful evidence here to help build your case.

Chapter 9 is for verification managers (perhaps you) – how can you effectively build and manage a formal verification team and what can you learn from the lessons of others?

Finally, chapter 10 talks about other formal applications you might find useful today or can look forward to in the (somewhat near) future.

We close with a few recommended books/papers you may want to read if you want to dig deeper and a glossary / magic decoder ring to help you with the sometimes-confusing terminology popular in formal circles.

Chapter 2 The Verification Treadmill

Why verification is so important

The goal of verification in semiconductor and system design is to prove that what we plan to build will do everything it is supposed to do and will never do anything it is not supposed to do. This is important in part because the cost to design and build one of these systems now runs to tens or even hundreds of millions of dollars; a trial-and-error approach to getting the design right would take an impossibly long time and become prohibitively expensive.

An even bigger concern for any enterprise is the possibility that customers might discover problems in their products. Issues can arise especially in use-cases that product designers didn't consider and therefore didn't cover in verification. For software, we're all too familiar with patch updates, but issues in hardware can't necessarily be patched; only a new chip can fix the problem. Making field changes to hardware is extremely difficult, in many cases close to impossible. Problems like this can have huge negative impact both on the supplier and their customers.

The ubiquity of electronics

Meantime, the complexity of electronics is accelerating rapidly. We now have high-resolution gaming, smartphones, LTE and 5G cellular communication, low-power design, cloud computing, semi-autonomous cars, smart homes, and the many other high-tech capabilities that surround our modern lives. These System on Chip (SoC) designs are in many ways significantly more complex than earlier systems, in size certainly (thousands of times larger than the Intel Pentium for example[2]). They also integrate more complex subunits such as multi-core CPUs, GPUs and other complex sub-functions, running multiple different types of software and inter-operating not only with each other but also communicating with the outside world over cellular, Wi-Fi and Bluetooth links. They're running much faster, with complex dynamic clocking and power management turning functions on and off in the middle of all this activity. Purely so you only need to recharge your device every few days.

Our tolerance to bugs is dropping. Where once problems in electronics were an inconvenience, fixable in the worst case by a reboot, now advanced systems control safety-critical functions in our cars or pacemakers or power plants. In these contexts, reboots are not an option and failures at minimum may lead to recalls, or worse still may cause fatal accidents. Security has become a major concern. The recently reported Meltdown and Spectre[3] bugs highlight how far we still have to go in containing security attacks. Where verification must try to find (and fix) every possible way in which a product might be compromised, attackers only have to find one way in and delight in finding obscure loopholes.

For all these reasons, product teams invest massively in design verification – at least 50% of the total effort that goes into

---

[2] http://www.wagnercg.com/Portals/0/FunStuff/AHistoryofMicroprocessorTransistorCount.pdf
[3] https://meltdownattack.com/

designing a product[4]. Thanks to hard work, clever techniques and continuing advances in verification tools, the industry has released many products which have worked and continue to work extremely well. But as design capabilities and demands continue to race ahead, it is inevitable that these verification strategies, comprehensive though they are, have started to show cracks.

The price of not being perfect

The earliest widely visible instance of a semiconductor verification failure in released products appeared in 1994 when a public post revealed that the (Intel) Pentium floating-point divide returned notably incorrect answers in a very small set of cases[5]. Intel have been verifying complex designs for a long time, they have a worldwide user-base depending on the accuracy of their platforms, they have accumulated massive test suites over years of development, and still a bug slipped through. Design and verification teams around the industry paid attention; if this could happen to Intel, who knew what unseen problems might be lurking in their own production designs?

$$\frac{4{,}195{,}835}{3{,}145{,}727} = 1.333739068902037589$$

*The Intel floating point bug –
the digits starting with 739 are incorrect*

Not finding these problems can be expensive. If you isolate a bug in-house after you manufacture (but haven't yet shipped) the device, you can do more testing but face potentially millions of

---

[4] There are differing views on this number – anywhere from around 50% to 70%. One interesting review is https://www.eetimes.com/is-verification-really-70-percent/
[5] https://dl.acm.org/doi/abs/10.1007/978-3-540-69850-0_8

dollars to fix the design. If the bug gets out into your customer base, costs explode. Intel reportedly took a pre-tax charge of $475M against earnings to correct their floating-point divide problem and update customers with the corrected device[6].

Safety and security considerations will further amplify the cost of bugs, perhaps as much in market impact and liability as in replacement costs. Media hair-trigger responses to bad news can drive instant drops in share price and may amplify reputational damage from which it can take years to recover. News of a glitch in an iPad app (used by pilots to access maps of airport runways) drove American Airlines stock down by $1.9 billion in the course of a few hours[7]. This problem was attributed to a software glitch, but we have already seen that hardware is not immune to bugs. Frankly, social media and markets don't care about that hardware/software distinction anyway. The tech failed in a serious way - dump the stock.

Why is this so hard?

At first glance, it might seem that we just need to verify harder or smarter, or maybe both. Unfortunately, no matter what we do, we can never ensure complete verification. It's important to understand why; this starts with how we verify.

The most popular method used in verification is simulation. We create and run (simulate) a series of tests and compare with the results we expect. When running a test returns the expected result, the test passes. When it doesn't there is a discrepancy between the design and the expected result, which may mean we have a bug in the design, or it may mean that our expectation was wrong. This approach, simulation-based testing through specific tests, often

---

[6] https://en.wikipedia.org/wiki/Pentium_FDIV_bug
[7] https://cdn2.hubspot.net/hubfs/69806/Reassessing_the_Cost_of_Software_Quality.pdf?t=1510935735043

called *directed testing*, is the natural way we approach verifying almost anything. It's also very effective, so much so that it continues to play a very major role in all verification today.

But it's incomplete. No matter how many tests you create and what clever tricks you use to cover multiple test scenarios in each test, you can only verify correct behavior across a finite number of possibilities, generally much smaller than the total set of possible behaviors.

This is an intrinsic problem in verification for any but the most trivial systems. To make this concrete, imagine the system is a phone and pushing a button on the screen starts a possible sequence of transitions between states which may go through thousands (or millions) of intermediate "next states" before finally delivering the expected outcome, starting a phone call.

Proving that this phone call always works correctly and never works incorrectly should test all of those possible sequences. If there were ten possible options (next states) at each stage and you wanted to exhaustively test a sequence of 100 steps (trivially short for hardware and software these days), you would have to test 10^{100} sequences, a task which would not be remotely possible even on a battalion of supercomputers[8]. The ***intrinsic*** complexity of verification grows exponentially fast with the number of states in the system (which in a hand-waving way is related to the size of the system) and with the length of the sequences[9].

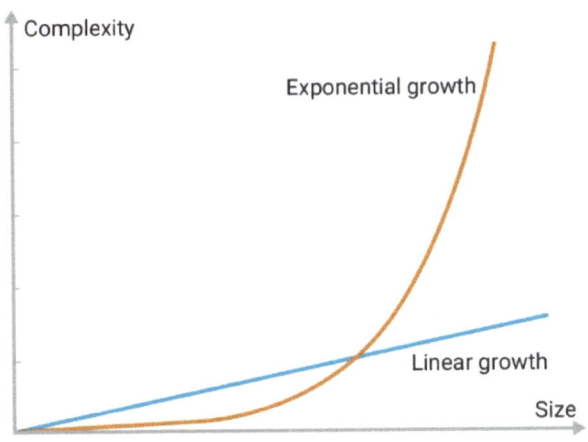

Exponential growth vs linear growth

Exponential growth is, with few exceptions[10], the fastest known growth in the natural world, and is much faster than our ability to speed up computers and software (Moore's law notwithstanding). Everyone involved in conceiving, building, and testing complex systems works hard to find ways to tame the implications of that growth for verification; clever techniques can often manage the

---

[8] If you could test a billion sequences in a nanosecond, you could test roughly 10^{25} in a year; 10^{100} would take 10^{75} years; roughly a trillion trillion trillion trillion trillion trillion years

[9] Even if you allow only two possibilities at each state, growth is still exponential. It starts slower but still exceeds practical reach very quickly. 2^{100} sequences might be practically checkable but a modest growth in sequence length gets you to 2^{1000}, which is again out of range of reasonable computation power.

[10] https://en.wikipedia.org/wiki/Hyperbolic_growth

early part of this growth to acceptable levels; as we'll see, formal methods have become important tools in this continuing battle. Still, it is important to remember that because of exponential growth, no one tool or methodology will ever become a long-term silver-bullet solution[11]. Verification will always depend on a range of tools and methodologies.

Battling the exponential

Because simulation is incomplete, a widely employed strategy is to decide when you have run **enough** tests. Verification teams get clever about this by testing bounding or corner cases which lie at the edges of acceptable behavior. For example, when testing arithmetic functions, using biggest possible numbers and smallest possible numbers as inputs is an obvious starting point for bounding tests. The reasoning is that if these extreme cases and a few other randomly sampled inputs in between verify correctly, all other cases should also work correctly. This tactic alone can massively reduce the number of tests required for that function.

Unfortunately, corner-case reasoning can be dangerous because it makes assumptions about the way the design is implemented, and those assumptions may not be valid. Often the design team decides that an architecture or implementation must be optimized for performance, size, power or other factors that may not be known to the verification team. In the case of the arithmetic example (say for multiplication) the implementation for small numbers may be quite different than that for larger numbers. Whenever there are architecture or implementation transitions like this, there are new possibilities for errors around those transitions. The verification team now must test not just at the extreme bounding cases but also at these new "architecture boundaries". Corner-case methods continue to be widely used, because we have no choice, but we

---

[11] Notwithstanding periodic debates, always entertaining, on formal verification eventually obsoleting simulation

must always acknowledge that there is an element of human judgement in the corners we pick.

A system bus managing data traffic between 2 IPs and a memory

Another challenge for directed testing is that sometimes the range of possible variations is simply too high for us to even think of all possible options, much less create tests for all those cases. A good example is a traffic manager (called an arbiter) on a bus through which multiple attached components (call them A and B) can communicate to a resource like a memory, but only one at a time .

When A is communicating with the memory, if B also wants to communicate it will make a request but must wait until the arbiter grants it permission. The arbiter may do this after A has completed, or if B has higher priority it may tell A to stop and allow B to start, but must eventually get back to letting A complete its task; it may even allow traffic from A and B to interleave in a controlled way. Since behavior depends on the order and sequence of requests to the arbiter (often from many more connected devices than we show here), relative request priorities, the amount of data to be communicated and what is pending in the request queue, the number of tests needed to prove correct behavior in all possible cases grows quickly[12].

Obviously, building tests to verify that even this basic arbiter works correctly across all possible combinations can be challenging. Adding real-world complications like interrupts, differing clock speeds, variations in architecture for IPs built by different teams, and many more factors, it quickly becomes impractical to build comprehensive suites of directed tests to cover all cases. Even corner-case tactics won't work here – there are simply too many corners.

Facing this problem with directed and corner-case based testing, verification engineers have turned increasingly to a technique known as *constrained-random* testing. In this method, they will build a test allowing for some aspects of the test to be randomized in a controlled way, those controls ensuring that the randomized test behavior remains reasonable. This technique in effect greatly expands the number of tests that can be run. One test-script spawns many tests, which can easily run in parallel by adding server capacity and simulation licenses.

Constrained-random testing has been very successful in teasing out potential hidden errors and is now a major component of any strategy in directed testing. But clever though this method is, coverage is still bounded by the number of constrained-random tests that verification engineers can write. While each script spawns many randomized tests around a particular objective, none are clever enough to run all possible tests.

---

[12] Assume 3 components attached to the bus (with 8 request scenarios – 000, 001, 011, etc), each with 3 possible lengths (short, medium, long) and each with 3 possible priorities (1,2,3). If the arbiter should remember up to 5 transaction requests for each function (which may come in any order), you need to consider 8x3x3x5x3 = 1080 possibilities. That's a lot of tests!

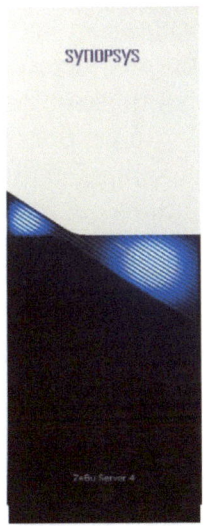

Synopsys ZeBu emulator

Another response to the scale problem is to use hardware-based acceleration, particularly emulation and FPGA prototyping. These technologies provide huge improvements in performance, running thousands or millions of times faster than simulation, which unquestionably helps a lot. Still, acceleration effectively offers only a constant improvement in performance (big though that constant is).

Reuse

Yet another approach depends on extensively leveraging proven reusable IP components in designs. If those IPs have been carefully tested and proven on multiple prior designs, the risk that they may exhibit problems in your design should be greatly reduced, at least in principle. This was another big step forward; however, reuse only provides confidence that those components work correctly standalone, as advertised by the vendor. There's no guarantee that *your* design will not introduce bugs in the way it interacts with those IP. Reuse **reduces** but does not **eliminate** the need for testing around those IPs.

All of these techniques are actively used today, but we still always come back to the exponential curve. No matter what we do, testing will never be exhaustive or anywhere near exhaustive because you can never test more than a finite number of sequences. The exponential growth in possibilities to test eventually dominates all these methods.

So product teams depend on expert verification engineers, designers and managers, who hold frequent reviews, track metrics like testing coverage and bug-trends, and who rely heavily on their experience and gut-feel[13] to decide when they have done "enough" testing. This process is solid enough that the semiconductor industry continues to ship successful products. But those darn designs keep getting bigger and more complicated and a question lingers. Beyond all the great testing that has been done, do hidden bugs still remain?

---

[13] It's not just about tools or metrics. Knowing that a particular function used in the current design has had a history of problems in previous products will alert an experienced manager to beef up testing on that function

Enter formal verification

A strength of simulation-based[14] verification is that it works naturally with the way we think about testing. We define a test, we write it, we define another test, we write that and so on. We always know how to further expand the range of tests we can supply. But we've seen the limits of this approach, not just in the sense of covering absolutely every possibility, but even in the sense of covering all the important possibilities.

A carefully designed and executed testplan should cover well all possibilities that we consider important in *normal* use, and also a set of *abnormal* use-cases that we deem possible. But what we consider "abnormal" is based on experience, subjective judgement and frankly, practicality. We have to put a bound on abnormal cases we are prepared to test to be able to complete verification in reasonable time. This can mean that we fail to consider unusual cases where a combination of conditions, building over many cycles, conspires to cause a seemingly impossible behavior, as in the *Meltdown* combination of speculative execution and cache behavior.

A different angle of attack seems to be called for. The problem with simulation is that we must handle testing case-by-case. We can test only at (carefully chosen) corners, we can cover more testing ground with constrained-random, we can run many cases in parallel or we can get big speed-ups through hardware acceleration. But all these methods expand testing capability by fixed amounts; none can overcome the exponential growth of inputs and sequences to be tested. We really need a method that can test all possible cases simultaneously (at least up to some

---

[14] From here forward we will use "simulation" to cover all the directed (and randomized) testing methods, including emulation and prototyping

point)[15]. We shouldn't forget also that we want to be able to do this at signoff quality for significant aspects of the verification plan; there's little added value in any technique which has only incidental impact.

One way to do this is to use variables for inputs and state values and a mathematical model for the design rather than the explicit states used in simulation. To illustrate, think of a 32-bit integer multiplier. In simulation, we test this works correctly by computing 1x2, 3x5, 7x13 and many other cases. Checking all possibilities requires 2^{64} tests (about 10^{19}) could take a long time. If instead, we could test with variable ("symbolic") inputs, say A and B, and mathematically verify that the output was always the formula A x B, we could completely prove correctness for all possibilities. This is the objective of formal verification.

---

[15] Perhaps quantum computing can help at some point, though there is no indication such a solution is near.

Chapter 3 Formal Verification – the Early Years

Background[16]

The principle behind formal verification is quite simple to state (though somewhat harder to implement) – turn what you want to verify into a mathematical proposition, then prove the correctness of that proposition. This is a very natural direction to take since digital designs are based on (Boolean) logic. You can think of a design as a (typically very complex) set of logical statements, and a behavior you want to verify (maybe "pushing this button will always initiate a phone call") as a mathematical theorem you want to prove in the context of those statements.

Premise #1: All men are mortal
Premise #2: Socrates is a man
Proposition: Socrates is mortal
Proof: Since Socrates is a man and all men are mortal, then Socrates is mortal.

Pythagoras Theorem Proof: $a^2 + b^2 = c^2$

Proof: Total square area is $(a+b)^2$
This area is also $c^2 + 4(1/2*a*b)$
Therefore: $(a+b)^2 = c^2 + 4(1/2*a*b)$
Equivalently: $a^2 + 2ab + b^2 = c^2 + 2ab$
And therefore: $a^2 + b^2 = c^2$

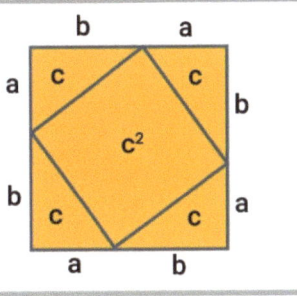

Mathematical proofs

---

[16] Drawn from A Brief History of Formal Methods

Mathematical proving has a very long and distinguished history, dating back to the earliest Greek philosophers who recognized that it was possible, given an appropriate construction of the problem and requirement, to prove statements which must be universally true. The Socrates example above is a very simple case illustrating the mechanics of a logic proof. You start with ***premises*** (which in our case correspond to the circuit description and perhaps some constraints on the allowed input behavior), you assert a ***proposition*** which you want to prove (in our case an expected behavior of the circuit), then you ***prove*** that proposition or theorem by following a logical and well-grounded sequence of steps.

We'll stress a point here because it underlies the basic advantage of formal methods. When you follow a mathematical approach, and prove a proposition formally, <u>*you have proved it (in a finite number of steps) for all possible cases*</u>. But when you simulate, you only prove for the cases you simulated; if you simulate a thousand cases but there are a million possibilities, you have still only proven a tiny fraction of what the formal method proved. This sounds so good that you might wonder why we still use simulation; it turns out that simulation and formal have complementary strengths (and challenges), as we'll see later.

Unsurprisingly, work in this direction advanced almost exclusively in academia and the big labs, in part because these were interesting technical questions but also because concerns were being raised in the US DoD and telecom companies, among others. Around the 1980s, a general sense that "we need to do better" transformed into more alarmed urgency, prompting active use for formal methods in several industries beyond the semiconductor ecosystem.

Boeing Dreamliner

In all cases, adoption of formal methods was prompted by publicly visible and serious failures on large and critical systems, including a radiation therapy machine delivering fatal overdoses of radiation[17], an Ariane rocket exploding 40 seconds into flight[18], Prius cars with an unexpected stall problem[19] and the recent discovery that the Boeing Dreamliner could lock up and lose control after 248 days of continuous operation[20]. Each failure was ultimately traced to rarely activated bugs which had been missed despite extensive testing yet had or could have had catastrophic consequences.

Just as we saw earlier for semiconductor design, a common conclusion from analysis of these problems was that dynamic verification (simulation) alone was insufficient to deliver high levels of confidence, especially in safety. Each of these systems providers enthusiastically embraced formal methods with an

---

[17] https://en.wikipedia.org/wiki/Therac-25
[18] https://en.wikipedia.org/wiki/Ariane_flight_V88
[19] http://articles.latimes.com/2014/feb/12/business/la-fi-prius-recall-20140213
[20] Boeing 787 Dreamliners contain a potentially catastrophic software bug

expectation that they could increase that confidence. In the early days, those companies, along with military, aerospace, telecom, and other system builders had to rely on custom-crafted tools adapted from university/lab research. Nevertheless, they proved that formal methods could be effective in proving correct operation, or in finding bugs that might otherwise have been very difficult to track down.

Now formal methods have been used to prove the correctness of driverless operation on one line of the Paris Metro[21] and operations of railway control systems[22]. Toyota applies formal analysis to prove correctness in a variety of car systems and Airbus has been using formal techniques for some time in validating correctness of avionics software[23]. In the very complex world of cloud services, Amazon Web Services (AWS) depends on formal methods[24] to prove correctness of operation of the various components of their solution.

Paris Metro

---

[21] https://en.wikipedia.org/wiki/Paris_M%C3%A9tro_Line_14
[22] https://www.prover.com/
[23] Formal Methods for Avionics Software Verification
[24] How Amazon Web Services Uses Formal Methods

Closer to home for readers of this book, Intel took the floating-point problem mentioned in the last chapter as a wake-up call to get serious about formal verification. Just before the turn of the millennium they used formal analysis to validate the Pentium-4[25], reporting that no problems escaped to the field matching the seriousness of the earlier bug. This industrial success, as much as technical advances in tools, contributed to growing interest in the field among semiconductor verification teams.

Since early formal methods software[26] was developed in academia and labs, only deep experts inside those domains knew how to run these tools. Over time, some of these experts migrated into commercial enterprises (such as Intel) where they started to build wider interest in these strange new techniques. But formal remained a highly-specialized art, barely impacting production design flows except in one immediately useful application requiring very little understanding of the underlying technology – logic equivalence checking.

Equivalence checking

When logic synthesis from RTL started to take off, an obvious question arose: how do I know the synthesis tool didn't make mistakes in converting from RTL to gates? When design sizes were relatively small, signoff verification[27] (mostly through simulation) was still common at gate-level so equivalence between the RTL and the gate-level implementation wasn't a primary concern – all that mattered was that the gate-level implementation behaved correctly. But as design sizes grew, high-coverage gate-level simulations became impractical; signoff verification increasingly moved to RTL, making the question of functional

---

[25] High Level Formal Verification of Next-Generation Microprocessors
[26] For example, SMV and ACL2
[27] *Functional* signoff verification, just to be clear

equivalence between the gate-level implementation and the RTL a much more pressing concern.

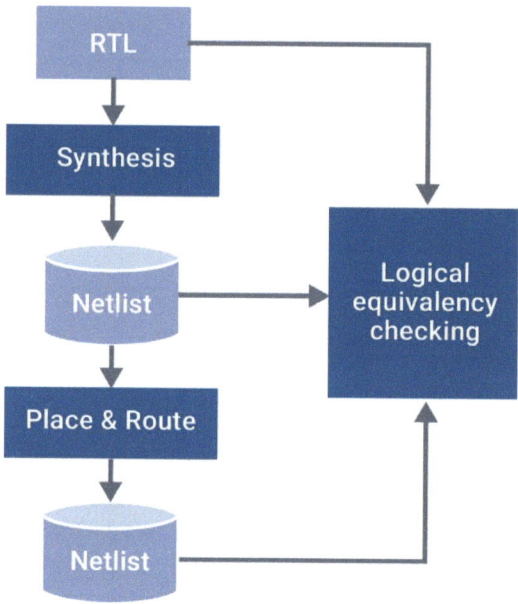

Equivalency checking

Addressing this need presents a perfect opportunity to apply formal methods. Any formal proof requires some kind of reference against which you're going to check. In this application, we have a ready-made reference – the RTL. We want to check that the **gate-level implementation** functionally matches the **RTL design**. There's no need to create additional statements about what should be checked, which simplifies usability[28]. Thanks to this ease of use and completeness in proving, logic equivalence checking has become a required signoff step in all major production design flows.

---

[28] In practice, in modern flows dealing with complex logic, equivalence signoff is still not completely pushbutton

Useful though equivalence checking application is, it still doesn't prove "correctness" in the more general sense we probably would like to see – correctness against intended behaviors of the design. Sure, the gate-level netlist matches the RTL, but how do we know the RTL is correct? Or that the architecture is correct? This requires a different kind of analysis.

What do we mean by correctness?

We all believe we know what it means for something to be correct, but correctness is one of those attributes that's not easy to define precisely. We tend to think the same way as Supreme Court Justice Potter Stewart who, when arguing about a definition of obscenity said that "I know it when I see it". This may work in legal decisions but is not useful for the kind of proofs we need.

US Supreme Court

When we take a mathematical approach, we need a precise specification of correctness. Here we run into a problem: what can we use as that specification? Perhaps we could start with the documented requirements? Every design, IP and block (at least for significant components) has some kind of specification, perhaps in

Word or PDF. But these are written in natural language (perhaps English) which rarely rises to mathematical precision:

> **English Statement**: Every access request is granted.
> **Possible meanings**:
> 1. Every access request is immediately granted
> 2. Every access request is immediately granted at the next positive edge of the clock
> 3. Every access request is immediately granted at the next positive edge of the clock, if the block is not in reset mode
> 4. Every access request is eventually granted
> 5. Every access request is granted after some allotted time
> 6. Every access request must be paired with a grant

Why English "specifications" struggle with precision

If documentation specifications won't work, perhaps we can use an architectural or SystemC or C model? Models of this type are sometimes available, but they are typically developed to explore and validate high-level features of the design; they are not normally defined with sufficient precision to act as references against which RTL equivalence can be checked. Think of a specification for an integer multiplier: "c = a * b". This may be sufficient for architectural modeling, but it doesn't specify timing (how many cycles are required to complete a multiplication) or power intent (how or when clocks or power should be gated), among other important factors.

In the few cases where these high-level models can be used as a reference (or extended to become an effective reference), then high-level equivalence-checking tools can verify the equivalency of the two models. Beyond datapath models, these opportunities tend to be uncommon and require significant investment in setting up and proving[29].

More commonly, whether starting from a natural language specification or an architectural model, design teams find that so much detail must be added to make the description sufficiently precise for formal proving that the effort required to build this reference outweighs the benefit[30].

> **Larger Ambitions**
>
> There is very active research in an area commonly called theorem-proving in which more extensive statements about what is expected of a system are proved through engines like ACL2, TLAPS, Coq, Isabelle and others. You might think of this as a more powerful use of property-checking requiring expert use, which can generate very powerful proofs.
>
> An unavoidable challenge in using this approach is that it requires a very detailed understanding and description of what constitutes correct behavior. Such a description may be at least as complex as the design RTL, both in intent and mathematical representation. Furthermore, theorem-proving is a still a mostly manual task for which an expert is required to drive the theorem-prover.

Theorem proving

---

[29] Another approach is to develop a high-level specification to be used for checking in a language such as TLA+. Both AWS and Microsoft Azure teams use this method in testing their software. Obviously, this approach requires further investment to learn and to build specifications and still requires an equivalence checking step for completeness (which is often skipped due to complexity)

[30] This is inevitably a balance between effort and economics or safety, which is why some verification efforts make the investment in comprehensive proving for some aspects of designs

A more practical approach to building a reference specification is to limit the scope of what we are trying to prove to local expectations of correctness, since these can be much easier to describe. Think again of our earlier bus-arbiter example; generating a full specification for such an arbiter could be extremely complicated, especially for complex bus protocols like the AMBA AXI protocol. Instead, you might in some cases choose to use simulation for some of your testing and formal for a **subset of behaviors** that are otherwise very difficult to test.

As one example, the arbiter must communicate with blocks through request and grant signals and will need to store pending requests in a FIFO. A very important specification is that this FIFO should never overflow, because if it does, requests/grants will be lost. Testing for this possibility could be very challenging in simulation because you must create heavy traffic/demand on the bus to overflow the FIFO. Even then you couldn't be sure that there might not be a case which would create such a problem among the many other possible configurations of traffic and demand that you hadn't tested.

A formal check for FIFO overflows addresses this concern; you can prove this specific problem can never happen (or isolate a case where it could happen). This approach, working with specifications which target important requirements within a function, is used in industrial flows today and is known as ***model-checking*** or ***property-checking***.

In certain cases, it is very possible to accumulate sets of property checks to provide a complete specification for a function, in which case formal verification can assume complete responsibility for the verifying the correctness of that function – no simulation required. This method is known as end-to-end (E2E) checking and is generally considered to be a fairly advanced use of formal.

Recapping, in common usage we abandoned hope of proving correctness against a complete specification and are now limiting ourselves to localized proofs of correctness, where correctness is expressed through properties, as we'll see next. In practice this is not a significant compromise; a complete specification may be redundant if simulation already provides much of the necessary confidence in the correct working of the function. Formal verification then complements this testing with targeted confidence for some especially challenging cases.

Properties: assertions and assumptions

Now we know we are checking properties, what are these properties? We'll start with assertions, which will lead us to properties. An assertion is just what it sounds like: "I assert that these two inputs can never be active at the same time", a behavior on which you depend but which may or may not be true in practice and which you therefore need to check.

The concept and practice of assertions was originally conceived[31] as a way to check *basic expectations* in software through executable checks embedded in the code; these would trigger automatically if requirements expressed through assertions were not met. An obvious example before a division operation would be an assertion that the divisor is not equal to zero, since division by zero is not defined. The intent is to catch basic problems quickly before they lead to later and more complex bad behaviors which might be more difficult to debug.

Assertions have been supported in hardware description languages for quite a long time, but widespread use of assertions as a part of hardware verification, known as assertion-based verification

---

[31] As early as the 1940's by Alan Turing
https://www.schematron.com/document/145.html?publicationid=

(ABV)[32], became popular in the early 2000s after standards emerged. From there, these evolved through OVL and PSL to the leading SVA (SystemVerilog Assertions[33]) standard of today. ABV continues to be very useful and popular in simulation; most importantly, for our purposes, formal tools adapted to read this same standard format[34].

In hardware design today, assertions are predominantly expressed in the SVA format; these can be embedded in the RTL for a design or can be provided through separate files. A simple assertion would be ***assert A == B***[35] which checks that signal A is ***always*** equal to signal B. If a formal tool proves this assertion, then the statement is true in all cases; conversely, if the tool finds this assertion is not correct, even for a single case, then it will report a case it has found where A is not equal to B (this is called a ***counter example*** or CEX).

OK, those are assertions but what are properties? A property is a formalized statement about the design with no attached expectations. For example, "my car drives forward" is a property I can associate with my car. It doesn't imply I can drive forward; that requires an assertion on the property. If I ***assert*** that "my car moves forward", now I am making a statement that it should move forward. If a property-checking tool could check this, it might report that "yes indeed, your car can move forward" or it might report that "no, your car cannot move forward because it's out of gas". In the example in the previous paragraph, ***A == B*** is a

---

[32] http://www.ijmetmr.com/oljuly2015/NKarthik-MGurunadhaBabu-MuniPraveenaRela-113.pdf
[33] SVA is a subset of the SystemVerilog standard
[34] Earlier, formal tools used languages like CTL, LTL and Sugar for property specification
[35] The actual SVA assertion is slightly more complex. We'll stick to this simplified form here

property and *assert A == B* is an *assertion*. An assertion checks the property following the **assert** keyword. This division of terms isn't nit-picking because properties can also be used in other contexts, as we'll see next[36].

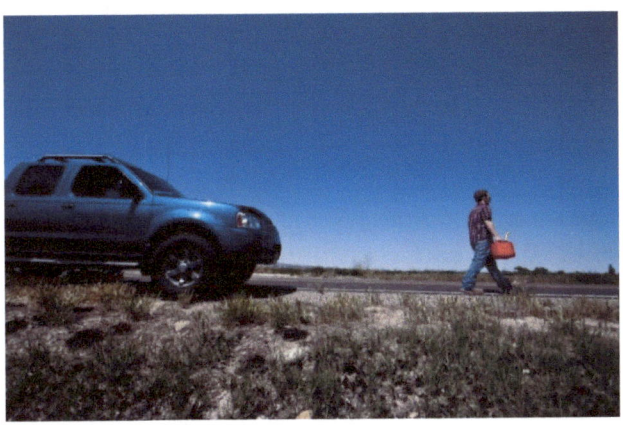

Sometimes it is necessary, as a part of proving an assertion, to constrain certain signals so that unreasonable or uninteresting possibilities are not considered. This can be done by defining *constraints*, which look very similar to assertions except that the keyword is *assume*. Using the same property, A == B, in this case *assume A == B* is an example constraint[37]. But where the corresponding assertion checks for cases where A is not equal to B, the constraint limits checking to just those cases where A is always equal to B.

In our car example, where the assertion is "does my car move forward?" a possible assumption could be "assume my car has

---

[36] In casual/common usage, *property* is often used as a synonym for *assertion*. Even the experts do this!

[37] Yes, this is a little confusing; is it an assumption or a constraint? When speaking about them, you can use either term, but **constraint** is most common. But the standard uses *assume*. Sorry, that's just the way it is.

gas". Together, the problem could be expressed as "assuming that my car has gas, can it move forward?" With this constraint, the formal tool could come back with "yes, the car can move forward" or maybe "no, it cannot move forward because the gearshift is broken" (isolating a more serious problem).

That's what assertions and constraints do; what do they really look like? Our goal isn't to help you write or even understand the detail behind properties, but it's worth knowing how to recognize the real thing in the wild. They can look rather complicated, but only the hands-on experts need to understand this stuff in detail. And actually, they really aren't as complicated as they first appear. Still, we don't want to scare you off, so we promise this is the only place you'll see these formats in this book.

```
Check that bus is a one-hot encoded signal (only one bit is high at a time)
bus_onehot : SVA: assert property (@posedge clk) disable iff(reset)
$onehot(bus) );

Check that when a request comes, an acknowledge should come within 5 cycles
SVA: req_ack_in5 : assert property ((@posedge clk) disable iff(reset)
$rose(req)) |-> ##[1:5] $rose(ack) );

Constraint that there request cannot remain high for more than 5 cycles
SVA: assume property ((@posedge clk) disable iff(reset) $rose(req) |-> ##[1:5]
$fell(req) );
```

Examples of SVA assertions

Now we know how to describe what to check (through assertions), we have to dive a little under the hood to understand how these properties/assertions are checked. We'll promise not to get too technical here.

Under the hood: the formal engines

At the core of any formal verification tool, you're going to find the model-checker. *Model-checker*[38] is just a fancy name for an engine that will take a circuit and a property (or rather an assertion) and will determine if that property will hold true in that circuit in all possible cases. If that doesn't hold, it should report a failing case, appropriately called a *counter example* (CEX).

To help understanding, we'll use a simple example – a familiar traffic light controller problem. In this case, we have two cross-streets with lights in both directions at the intersection. Lights in each direction can cycle through red, yellow, and green states. The obvious safety property[39] we want to check is that we can't get green in both directions at the same time. If this should fail to happen under any circumstance, the outcome could be catastrophic. Checking this requirement is an excellent application for formal.

---

[38] Ed Clarke pioneered model checking especially while at CMU and has received multiple awards for his work in this area.
[39] A safety property is a property which must always hold true, popularly summarized as "nothing bad happens".

Think of a model for this controller based on two finite-state machines (FSM), one for each direction, describing when and how the lights can change. These FSMs need to negotiate to determine which set of lights is going to change to what state (red, yellow, or green) next. Each FSM has to consider not only its own current state and next possible state given various inputs (e.g., is there a car stopped at my light) but also the current state of the other FSM. You should see now that it is tricky to know for certain that the safety property will never be violated, especially when there may be yet more inputs like pedestrian crossing requests. This type of property-checking is also very relevant in SoC bus design where only two devices can be allowed to communicate through a common bus at any time.

Traffic lights also provide a good example for liveness checking[40]. A light might avoid violating safety checks by never turning green, but this is also not desirable behavior, at least for drivers stuck on red. We need to add a property check that each light will turn green eventually (within some acceptable limit in practice). Similar conditions apply in SoCs, again especially around bus communication. An IP wanting to communicate through the bus should not be stalled indefinitely (often associated with hangs or deadlocks)[41]. Each should get a chance to communicate no matter what other demands there might be on the bus.

---

[40] A liveness property is a property which should eventually be true, popularly summarized as "eventually something good happens"

[41] There are very real problems SoC designers watch out for in these cases: **deadlock** where control is stuck in one state and nothing happens, **starvation** where one resource is blocked from access while others continue to have access and **livelock** where two or more resources are locked in a struggle for control and still nothing useful happens!

We don't need to discuss here how the controller is designed, only how we are going to check those properties we specify. Remember that the simulation approach to exhaustively verify the design would be to cycle through all possible input and state combinations, and to check that no assertions fail. In a formal approach, instead we will calculate with variables in place of those explicit values and we will use mathematical techniques to reach a conclusive proof. Here rather than simulating, we build equations expressed in Boolean logic form; these equations cover all possible values, so if we can prove our properties must be true given this set of equations, we have proved it in all possible cases. This approach, called **Model Checking**, was the first big step towards property checking in hardware verification.

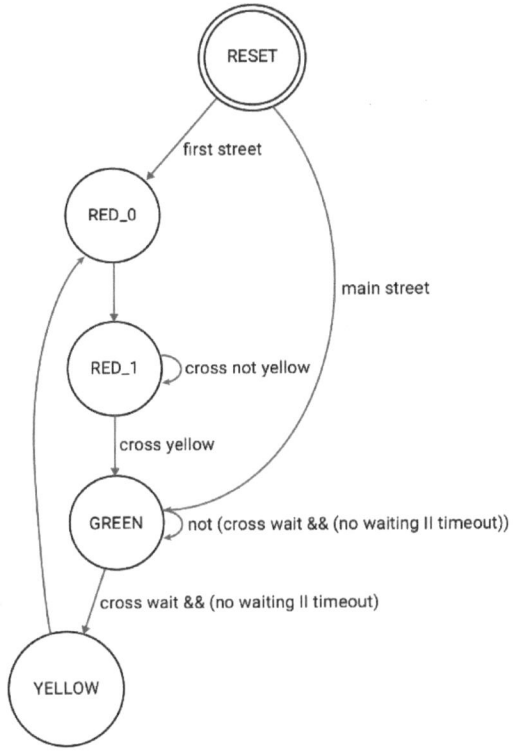

An example finite-state machine (FSM)

The details of how this is done are too technical for our purposes, so we'll attempt a *very* simplified explanation. Any set of Boolean equations (and therefore any digital logic circuit) can be represented as one or more interacting FSMs, which can be graphed as a set of states (the bubbles in the picture above) and possible transitions between those states (the arrows). Our traffic-light controller can be graphed in this way. This graph can in turn be mapped to a different type of graph which is designed to be more efficient for proving properties.

The flow to accomplish all of this starts with the front-end of a logic synthesis flow, which maps your RTL into a control/dataflow graph, from which it builds and optimizes one of those specialized graphs. This correspondence between logic synthesis and formal verification shouldn't be too surprising; the Berkeley ABC[42] platform is a widely used and adapted platform of this type, expressly designed to serve the needs of both synthesis and verification.

Optimized graphs come in different flavors, depending on the engine. One style is the binary decision diagram[43] (BDD). Model checking based on BDDs, called Symbolic Model Checking (SMC), became the forerunner of all modern model checkers and made possible property-proving for designs with 100-200 flip-flops. Impressive progress, but hardly up to the needs of modern designs or even sub-functions. Naturally this triggered more research, to the point that BDD-based SMC methods can now handle designs up to thousands of flip-flops (if components like memories are abstracted out in some manner).

While SMC with BDD showed promise, capacity was still a major concern. BDD memory consumption grows exponentially with the

---

[42] http://people.eecs.berkeley.edu/~alanmi/abc/
[43] https://www.cs.cmu.edu/~emc/15414-f12/lecture/bdd.pdf

number of states in the circuit, spectacularly shooting to sizes over 4GB in a matter of seconds, making the approach impractical for many real problems. Looking for a different approach that didn't so quickly succumb to unusable growth led researchers to **Bounded Model Checking (BMC)**, using a proof method called **Boolean Satisfiability Solvers, or SAT Solvers**[44].

```
x = b OR c
y = !a OR !d
z = !b OR d
p = !(x AND y AND z)

One SATisfying assignment for p to be
false (0) is:
         a=0, b=1, c=0, d=1
```

An example of SAT on a simple logic design
SAT tests if property p can ever be false; in this case it can

Instead of building the complete problem representation required in BDDs, BMC+SAT switched to a breadth-first approach, looking for a violation of the property to be checked (a counter example) within a pre-determined bound on clock-cycle depth.

Bounding the depth to which the search extends can significantly reduce the size of the analysis problem, making proofs (and especially finding counter examples) much more feasible both in memory requirements and in time. At the same time, BMC naturally handled sequential behavior by *unrolling* sequences. The

---

[44] SAT techniques have been around for a long time, getting their start in artificial intelligence for applications in planning / scheduling

next cycle in logic (for all possibilities beyond this cycle) is expanded as a new set of logic, taking the previous cycle states as inputs. And so on for continuing cycles, out to whatever bound is set. Analysis can then just work with this sequence of logic stages without having to worry about clock cycles.

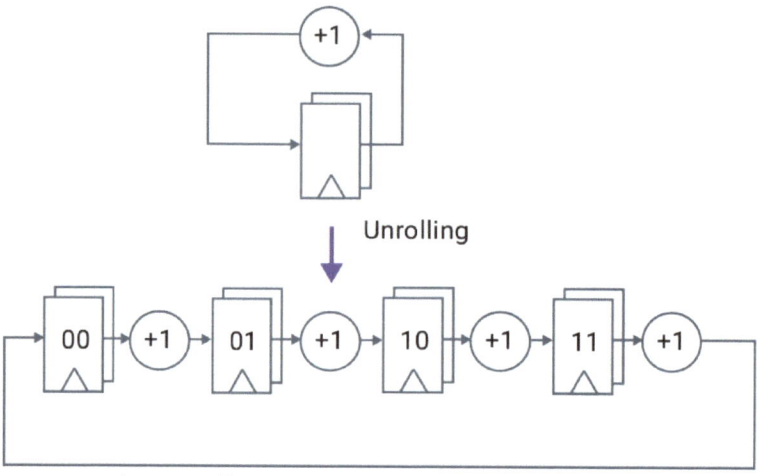

An unrolling operation on a 2-bit counter

The boundedness of BMC working together with the natural solution-finding power of SAT has proven very successful, so much so that now BMC+SAT is now one of the dominant approaches to model-checking. But it's not perfect; while bounded methods will often satisfactorily conclude proof of a property, they can in some cases fail to prove that property or fail to find a counter example within the bound that has been set. In this case, the result is ***inconclusive***. We'll talk about this more in the next chapter. Inconclusives (also known as bounded or undetermined proofs) are an unavoidable feature of all formal methods but, as we'll see later, they don't have to be dead ends.

Tools and methods have continued to evolve at a rapid pace so now there is quite a range of engines, techniques and flows to formally attack a property-checking problem. Exploring all of these would take us too far from our goal of providing an introduction to the field. If you want to learn more about this rich set of possibilities, check out our suggested reading list at the end of this book. And remember that innovation in formal methods hasn't stopped; you should expect to see yet more capabilities appearing in production tools[45] [46].

Among this range of proving engines, each engine has strengths in addressing certain problems and weaknesses when facing others. For this reason, we need access to a toolkit of methods to attack the wide range of problems that will arise in real circuits. Managing our way through these options is a topic for our next chapter.

---

[45] http://fmv.jku.at/papers/prasadbieregupta-sttt-7-2-2005.pdf
[46] http://www.springer.com/us/book/9780387691664

Chapter 4 What's the Catch?

Property checking goes commercial

Thanks to the promise of property checking and success in some high-profile design companies, several commercial products started to appear around the early 2000's, some from the larger EDA vendors, others from new ventures, and were actively promoted as a new direction in verification.

In each case, leading edge verification teams were enthusiastic, using these tools primarily to address the hardest problems that were proving intractable for dynamic verification. Naturally, everyone assumed that over time more verification teams would become comfortable with formal methods and adoption would quickly spread. Some enthusiasts even hoped that formal verification might eventually replace simulation. But it didn't work out that way. Why it didn't might be attributed to several factors and is the subject of this chapter.

A problem with assertions

The idea of adding assertions to a design is simple; the practice of adding effective assertions is not always so simple. The easy cases (a queue should never overflow for example) represent a small fraction of the cases you really ought to check. And putting lots of easy assertions throughout the design isn't generally very useful. More important usually is to check more complex bounding cases dependent on multiple states and tricky sequences, one common example being checking for correct interface behavior between blocks in the design.

In writing the associated assertions (and assumptions), you have to think hard about the theoretical operating bounds of the design in order to correctly draw the line between legal and illegal operation. Draw this line in the wrong place and you may report errors on operations which are legal, or you may fail to error on operations which are truly incorrect. Getting this right can be quite tricky since you have to imagine the limits of legal use cases, whereas in simulation you just run tests to see if any bugs appear[47].

Many verification teams found the investment they had to put into thinking of, creating and debugging high-value assertions for formal verification was sufficiently onerous[48] that they would build some, but overall assertion density (as a measure of how effectively you were using assertion-based verification) was not very high. Highlighting this problem, checking the correct behavior of an interface IP (requiring assertions running to hundreds of lines) would have been far out of reach for a typical verification team. And where teams had already adopted assertion-based verification (ABV) for simulation methodology, they often reported that adding constraints ("assumes") to correctly bound proofs took as much effort (or more) than developing the assertions.

Completeness is expensive

Remember that formal methods work by analyzing a symbolic model of the logic, rather like solving algebraic equations generally. This has the great advantage that proofs (or bugs), when found, are certain; there's no need to test additional cases. But you pay a price for this completeness. Problem complexity in formal

---

[47] OK – simulation teams work hard to find those cases too. But absolute proofs should be the central value of formal, so "best efforts" don't really measure up

[48] The effort required per assertion (or group of related assertions) isn't abnormally high. It is often on the same order as building UVM stimulus generators and monitors/checkers. But it's added effort which must be traded off against other verification investments

methods grows exponentially with circuit size, even with the most powerful formal engines; in fact, there is no guarantee that any given problem can be solved in a finite run-time on any size machine[49].

This is not the only problem in design engineering which is theoretically unbounded. Place and route (particularly routing) is an instance of the travelling salesman problem[50] which also theoretically may never complete in reasonable time/space bounds. Yet place and route is absolutely routine in digital design today. Formal methods have a similar limitation but continue to be useful in finding difficult bugs beyond the reach of simulation-based methods. In both these applications, what could in principle be impossible has been wrestled into practical usefulness in most cases through significant advances in empirically discovered best-practices. But we should remember that completion is not guaranteed; some cases may still require impractical or even unbounded run-times or memory.

Since formal methods are bounded, if the problem space becomes too big you could just surrender. But sometimes it is worth switching to a different formal method because maybe that proof or bug you are looking for is just a little further out. Remember all those different engines and techniques we talked about in the last chapter? Tricks to see if it might be possible to complete a proof are to try a different engine or to try the same engine with different parameters. Or you might try decomposing the problem into smaller pieces which may be easier to solve separately. In fact, multiple techniques can be applied. Managing all of these options starts to require more expertise on the part of the verifier, which becomes more apparent in the next section.

---

[49] Famously proved by Alan Turing in 1937: https://en.wikipedia.org/wiki/Turing%27s_proof
[50] https://en.wikipedia.org/wiki/Travelling_salesman_problem

Inconclusive and debug

If during proving the formal tool stops, it can report one of three possibilities for each property: that a property has been proven, or that a counter example has been found (maybe a real bug or an artefact of insufficiently considered constraints), or in some cases that the run was unable to run to completion beyond an acceptable bound. Outcomes of the last type are known as *inconclusive* and happen when the tool exceeds a set bound in memory or time. Then you must consider your options. A few possibilities were mentioned above – use a different engine or change parameters for the engine. Another option is to manually guide the flow of proving through various methods.

A common way to reduce the size of proofs is to replace an embedded block of functionality with a simpler model covering only what you believe to be the most important behaviors, a process known as *abstraction*[51]. This may be as simple as replacing a block with a black box, if that functionality isn't important to what you want to prove. Going back to our earlier car example, if we want to prove that a car can move forward, we don't need to worry about the details of windows, windshield wipers, infotainment and so on. We can start with an abstracted model of the car with only the engine, wheels, and drivetrain. We can't model the transmission as a black box, but we might abstract to a simpler model, considering only the *park* and *drive* states and ignoring *neutral, low,* and *reverse* options.

Abstracted car

---

[51] The Art of Abstraction

In the design world, we might model a memory as a black box, effectively allowing for any possibilities and sequences in address and data behavior[52]. Conversely, counters can be challenging for provers because these have also many possible states, but a black box model may be too unrealistic. Instead, you'll typically abstract the counter with a greatly simplified model which maybe considers only values significant to downstream logic. For example, if a downstream FSM changes at counter values 2, 5 and 10, only these three values may need be modeled in the counter; all other possibilities are collapsed into a default case. Proving then has a much smaller state space to handle and has a higher chance of completion. But doing this correctly isn't trivial; you must reduce the state space enough to enable completion but not so much that you may miss real problems.

Assume guarantee

assume always !(A & B);

Block X → A, B → Block Y → assert...

assert always !(A & B);

Assume-guarantee applied between two blocks in a proof

Another approach is to decompose the problem into smaller parts as shown here and use a technique called **assume-guarantee** at the interface between those blocks. Since each part is smaller than the original problem, property checking in those independent parts is more likely to complete successfully. The blocks are connected through properties which are used as constraints (**assumptions**) on

---

[52] Which might be an OK choice in some cases and not OK in other cases – depending on use-model

the inputs to the downstream block. Those assumptions are in turn verified (***guaranteed***) to hold at the outputs of the upstream block. Careful use of this technique can reduce problems which are unsolvable, or which complete only in hours, to a set of sub-problems each of which can be proven in minutes. But of course, you have to figure out how best to divide the problem and what invariant properties you will use at each interface. You also need to make sure the way you divide doesn't generate false problems. All manageable, but this is not for beginners.

There's yet another approach to manage difficult proving problems, using constraints. These will limit the scope of a proof by forcing certain signals to take a limited range of values. For example, a USB IP may be configurable to run in 32-bit mode or 64-bit mode. Either mode is legal, but the IP may only be used (at any one time) in one of these modes. A formal tool won't figure this out on its own; you have to specify a constraint. If you don't, it is quite possible that you will get an inconclusive result or perhaps a meaningless counter example reflecting unrealistic usage. This technique, systematically splitting a problem into separate use-cases is known as ***case-splitting***. Often the cases are fairly obvious, but you have to be certain or somehow prove (perhaps using assume-guarantee methods) that there is no possible interaction between operation in different cases.

Stuck at the chasm

You might now have a sense of why formal verification didn't instantly spread everywhere. Where it works, it works very well. But in many cases, getting it there can take quite a lot of expert supervision and effort. Those experts were able to figure out which proof engines to use and with what parameters when something got stuck. And they knew how to apply appropriate guidance to the tool to confidently validate behavior without hiding problems.

Design teams and tool vendors quickly learned that the most successful way to deploy formal tools was to build an army of formal experts and farm out all the formal problems to them. Some of the tool companies followed suit, building teams of highly expert AEs, many with advanced degrees in formal verification, who would work closely with customers, in many cases running the tools for them. The need for help also prompted new companies who specialized in consulting for formal applications.

This service-intensive approach worked but necessarily limited scaling use of the technology. Formal verification couldn't expand to being used widely because there weren't enough experts available, and even if more experts could be trained, that service-based use-model would be too expensive to scale in the long-term. In the terminology of Geoffrey Moore's book ***Crossing the Chasm***[53], formal verification was stuck on the left side of the chasm. The experts on the left side (small teams in perhaps ten large companies) were happily using formal in expert use-models, but there was no way this kind of usage could cross over to the larger market and mass adoption. Something had to change.

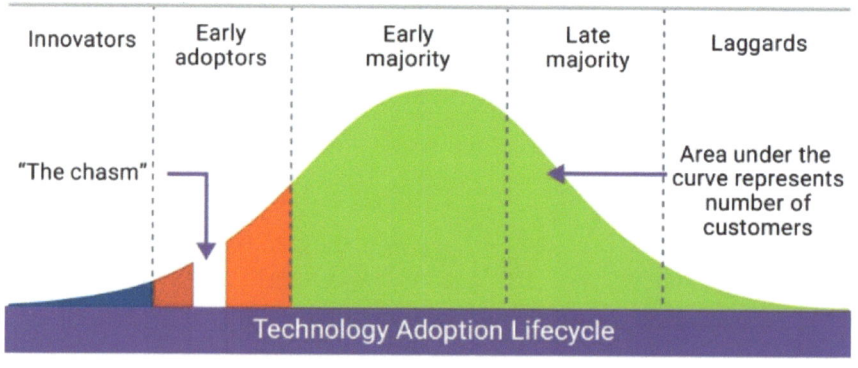

Crossing the chasm

---

[53] https://en.wikipedia.org/wiki/Crossing_the_Chasm

Chapter 5 Crossing the Chasm – Fearless Formal

Pre-packaged solutions

To recap, sometimes formal methods would find critical problems, but sometimes they wouldn't, or couldn't deliver a useful result without additional complex effort. The return on investment was uncertain for many verification managers, indeed even for the companies supplying these tools.

Pre-packaged solutions seemed like an obvious answer. We have verification IP (VIP)[54] for simulation, why not also for formal? Assertion IP (AIP) (also known as assertion-based VIP or AB-VIP) are indeed a part of the answer and vendors offer solutions for a range of interfaces. The same concept is also scaled down to other simpler yet still widely used design components such as FIFOs, linked lists and special CDC synchronizers, where assertions/constraints can be packaged with that design IP.

Over time, formal experts found a complementary approach in what is now often known as design or verification *patterns*. These aren't associated with blocks in a design necessarily but rather with commonly occurring verification objectives and processes. Each time the formal team would address a certain type of verification objective, say checking clock connectivity at the SoC top-level, they found they were building similar scripts, similar assertions and constraints and applying similar abstractions, even following similar paths in decomposing large problems.

---

[54] A verification IP is a model used in place of a design IP (such as a USB function) to verify correct interaction of the rest of an SoC design with this IP. VIP are heavily tested against associated standards and are widely used for their reference quality. Simulation VIP also include verification support through assertions, debug support and cover properties

Whenever patterns emerge, that's a strong hint that it should be possible to provide value in a different type of automation – an ***application*** or ***App***. Pre-canned scripts, assertions, etc. aren't sufficient to handle many possible SoCs with different architectures and objectives, but a combination of a greatly simplified user input along with application-specific code can, under the hood, construct and drive all the steps in a specific pattern. This includes not only problem setup but also running and managing run-time issues through all the methods we described earlier. You'll even find app-specific debug in many cases. The app approach really caused formal adoption to take off, so much so that today you'll find 10 or more apps offered with each of the major formal platforms.

Incidentally the app-based approach doesn't eliminate conventional property verification. It is still possible in app-based systems to check your own custom properties. There's generally even an app to help you do that!

Formal apps through the SoC development flow

A possible misconception about apps is to view them as the beginner's version of formal, to be abandoned as soon as you have built enough expertise. That view is not correct. While apps simplify use of formal methods in their target objectives, they are not verification lightweights. Even advanced verification teams continue to address high-value problems through apps, much more effectively than they could through other verification flows. In fact, it is not unreasonable to expect that over time more standard patterns will emerge and be handled through yet more apps. It is arguably better to consider Apps (and AIP) as the backbone of formal application, with custom property verification reserved for those cases not yet covered by packaged solutions.

Since most verification (and some design) teams get their start in formal through these apps, we'll discuss a few of these in some detail. To avoid confusing generalizations across different products we'll use Synopsys VC Formal and a few of its associated apps, starting with the apps that are easiest to adopt. You should remember that products and apps from other vendors may differ in some features and/or use models.

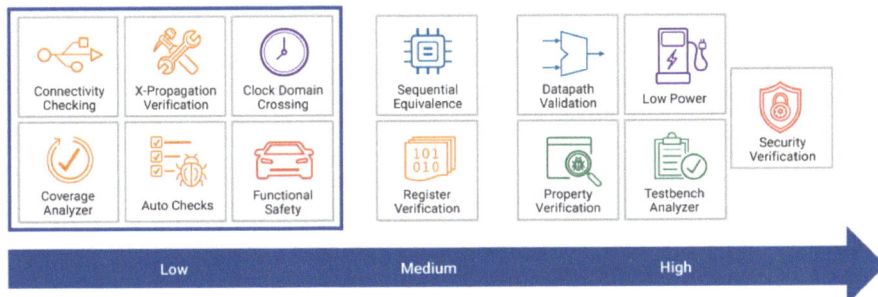

Apps ordered by ease of adoption
In this chapter we review the lower effort apps on the left

Auto checks – functional linting

The simplest app goes by a few names – auto-extracted properties (AEP in Synopsys VC Formal) or functional lint are examples. This app looks at the RTL for a module or block, generating several assertions representing standard "best design practices" which are then checked automatically. You never have to be concerned with the internals of those properties. Some of these checks are often associated with linting, but the formal versions checked in the app are less "noisy" (report less false errors) than you would find in a pure lint check. Most important for those who want to get started with formal, running these checks is almost[55] as simple as running linting. As a bonus, when an issue is found, it is accompanied by a waveform, so it is easy to understand the problem.

```
reg [7:0] a,b,c;

...
assign a = a & 63;
assign b = b & 127;
assign c = a + b;   // a typical linter would flag this as an overflow hazard
```

Apparent lint problem which is not a real problem in this case

Take for example an arithmetic overflow check. Suppose the RTL code adds two eight-bit (unsigned) numbers and puts the result into an eight-bit (unsigned) register. It is possible this value can overflow the result, which is what a lint-check would report. If you add 16 (decimal) to 240 (decimal), the result is 256 which requires 9 bits and therefore overflows the 8-bit result. However, if the first number is in practice limited to never be bigger than 63 and the second number is similarly limited to never be bigger than 127, no error should be reported. Proofs in this case require formal analysis to determine the functionally possible bound on the total sum. Here, the formal check will be less noisy than the lint check, minimizing engineering effort to check false errors which makes this app popular in design teams.

There are other checks in the AEP app, for example checking that a high-impedance bus can never have more than one active driver and that it can never be floating, or that a finite-state machine (FSM) has no inaccessible states (states that cannot be reached through transitions from other states in the FSM). The main point

---

[55] You may need to add some constraints in some cases

about these checks is that they are really as simple to run as lint, and any counter examples (violations) will be reported in the standard verification debug window. This starting step into formal is so easy that anyone who understands RTL can use it without any understanding of formal verification.

X-propagation verification

In simulation, 'X' is used to indicate an unknown state (which might be 0 or 1 or some transient state). Registers which are not reset on design initialization will start in this unknown state. In some cases, this is not a problem; the state is set to a known value before the value is needed. In other cases, if this state is not reset an unknown value can cascade through subsequent logic causing serious misbehavior. X-propagation analysis looks for and flags all potential problems by injecting Xs in vulnerable logic and automates generating assertions at observation points. The figure below shows a typical flow.

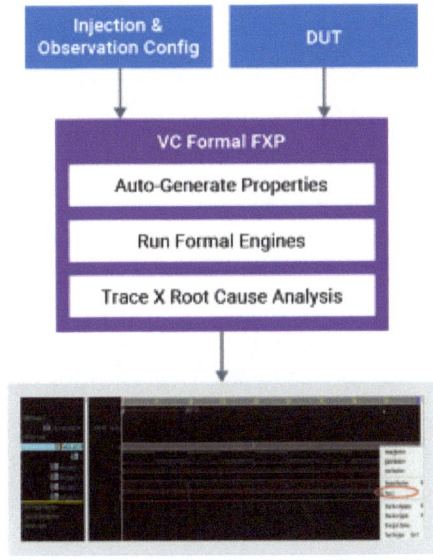

X-propagation analysis

Coverage analyzer – formal helping simulation

The later stages of coverage-driven signoff are always painful. As you build and run test-cases, initially coverage rises steeply. But the more you progress, the harder it becomes to increase coverage. You keep adding more tests after carefully studying which parts of the design are not yet being hit in testing, yet each new test barely moves the coverage needle, if it moves at all.

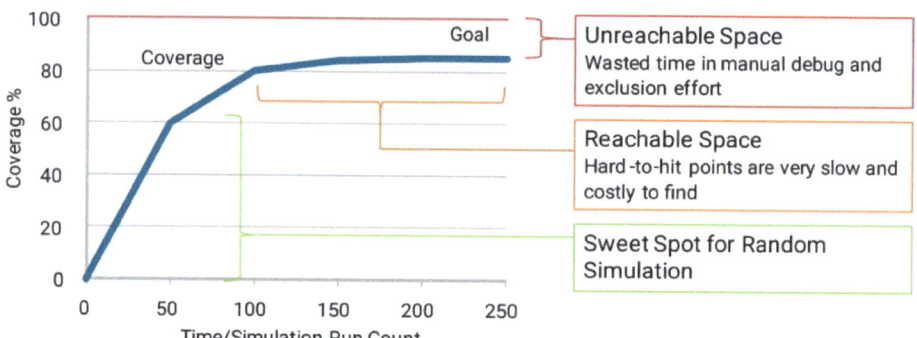

Progress and challenges in coverage closure

An important part of this problem is that, usually, some parts of what seems should be covered simply cannot be covered by any test – they are inaccessible or, in formal terms, unreachable (known as UNR in the simulation world). This might seem odd – if RTL code is not used, why not get rid of it? Because it might be inside a piece of logic which is used in some designs, but not in your design. Or it might be legacy code, lingering in the design because no-one is really sure if it can be removed safely. Remember the engineer's maxim: if it ain't broke, don't fix it, because it's quite likely you'll break something else if you try! So you leave it in there, but you can't ever get coverage on that code in your design and you don't really know what is truly unreachable. Or what should be reachable if you could only find the right test. Figuring this out is part of what takes so long to get to coverage closure.

This is where a formal app can help – the Formal Coverage Analyzer (FCA) in the case of Synopsys VC Formal. Formal coverage analysis works together with your simulation environment (with or without existing coverage results) to find logic that is provably unreachable. You exclude this logic from subsequent simulation coverage runs, giving your simulation team a better sense of how much testing they really have left to do. This app alone can have a huge impact on conventional verification effort and schedules.

Formal coverage analysis requires little to no additional input beyond the source RTL. If you already have simulation data, you can use it to focus attention on what has not yet been reached. And you can make your analysis sensitive to all the standard coverage metrics when building unreachability lists: line, condition, toggle, and FSM. The Synopsys VC Formal App can be even turned-on during simulation through Synopsys VCS. Which means most verification engineers are not even aware that formal analysis is being run; unreachable states are simply removed from coverage analysis. How cool is that?

Connectivity checking

In modern SoCs, most of the clever functionality has moved to (reusable) IPs/blocks – CPUs, GPUs, peripherals, sensor management, DSPs, and others. At the top integration level of the chip, design is reduced (almost) to simply connecting all these functions, and this creates a new verification challenge. We mentioned earlier that reuse emphasizes extensive verification at the IP/block level so that functionality doesn't need to be re-verified at the SoC level. But then how do you verify all the top-level connectivity is correct?

A block diagram of a typical complex SoC
Each block may have thousands of connections at this level

There are massive levels of connectivity in SoC integration (tens to hundreds of thousands of connections) and, despite that goal of moving all the cleverness into IPs, still quite a lot of logic must be generated to fully manage integration. This includes bus management logic, power management logic, debug, test, and IO control logic, and often security and safety management logic. Much of this is created quite late in the design schedule, simply because some of these features can't be completely finalized until the rest of the design is finalized. That's not good news for simulation-based checking which generally requires significant effort to build testbenches; if that effort can't start until the rest of the design is complete, schedules stretch further out.

Certain specialized simulation testbenches can be used to check connectivity (and often are for controls like clock and reset) but these approaches become increasingly difficult to manage as higher levels of configuration control are added. In structures like

IO connectivity with many complex configurations, getting to satisfactory coverage can become even more challenging.

A simpler approach is to specify and check this top-level connectivity through a connectivity check (CC) app. For point-to-point connections checks (is this output pin connected to these input pins?) – each connection can be represented to the tool as a single command. Lots of connections at the top-level are like this and can be checked using these simple commands – often by extracting a list of connections from an early revision of the top-level RTL, manually checking that list then using it to regress later revisions[56].

Some of the connectivity required at the SoC top-level is more complex than point-to-point connections; one example is the input/output (IO) logic. Most SoCs have many more internal IOs than can be supported by the limited number of pins on a practical chip package. However, all functions in the SoC typically don't have to be active at the same time so can be managed by multiplexing signals between different function blocks and the IO pads. Architects or application engineers build a complex spreadsheet defining which signals should be accessible at the same time and under what conditions so that all internal functions can be accessed under appropriate settings. From this spreadsheet, scripts automatically synthesize the muxing logic which will manage connectivity between the core and the IO pads.

---

[56] A popular intermediate format for handling large numbers of connections is a file of comma-separated values (CSV), which can be viewed, checked and corrected in a spreadsheet tool.

Input path I/O logic - The mux controls determine whether the pad or another internal signal drives Model Y

Checking this logic requires more than point-to-point verification, but these structures are still relatively simple, which makes them very easy to check in formal verification. IO muxing is completely combinational (if you don't include the registers controlling mux selects) and is typically not very deep so you don't have to worry about hitting verification bounds. Assertion generation can be scripted easily from the same spreadsheet, or through a different approach if you want complete independence between generation and verification. And a formal proof will check every possible variation of signal accesses between pads and the core logic under all possible mux control settings, something which would require considerable effort in simulation setup.

While in the IO example logic is predominantly combinational, this isn't always the case for logic generated at the top-level. Reset trees in an SoC can be quite complex and may include registers. Debug bus muxing may also include registers, either to hold snapshot values or to help in sequencing wide-debug snapshots onto a narrower debug bus. Even so, these structures remain relatively simple and are well-suited to formal proving.

Setup for connectivity checking is a bit more involved than for AEP, but still requires no formal expertise. You must create scripts or spreadsheets to define the connection checks, one-line checks for point-to-point connectivity and somewhat more complex specifications for structures like IO muxing or debug busses. In almost all cases, verification teams find ways to generate these scripts from existing files – from a top-level RTL or spreadsheet for point-to-point connections, or IP-XACT files, or from legacy data or specification spreadsheets for structures like IO muxing. Naturally this implies need to verify the scripting and the first-pass spreadsheets are correct, but many teams still claim high value and total engineering effort saved in this checking, especially in regression, even for SoCs as small as 20M gates.

Clock domain crossing

SoC design and CDC analysis go hand-in-hand. Any device supporting (at minimum) multiple peripheral interfaces must support multiple clock domains, implying lots of clock domain crossings and need for care in managing metastable states and lost data at those crossings. Many design or verification teams look only at structural analysis of these crossings; that analysis is very important but can be further enhanced with functionally aware analysis.

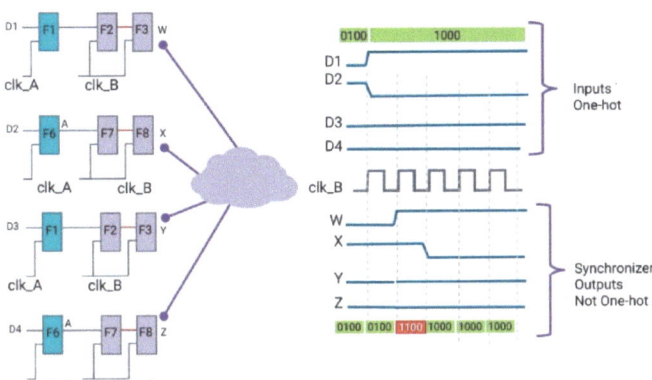

When multiple paths between two clock domains converge they must be one-hot encoded to guarantee correct operation

One example check considers the correctness of data transmission between domains. Conventional synchronization deals with metastability at crossings, ensuring that a register at a crossing doesn't lockup under certain conditions. However, this alone doesn't ensure that transitions may not be dropped or shifted in a crossing (for single bits) or may not become temporarily invalid (for vector signals).

Handling such cases requires careful design; handshakes, grey-coding or one-hot coding are methods commonly used here. Formal verification is a great way to fully check that this crossing-management logic has been implemented correctly. Conversely, engineers may define waivers to suppress reported "errors" they believe are in fact OK. A formal app can check these waivers for correctness. The more advanced CDC tools will offer these kinds of checks alongside structural checks. They may also offer false-path and multicycle path checks, valuable in CDC analysis, also in timing analysis.

Functional safety

Safety verification is another domain which is very hot, especially around automotive applications. An automotive risk classification (ASIL rating) defined within the widely adopted ISO 26262 standard determines in what applications a component can be used. The most safety-critical functions require an ASIL-D rating. A part of determining this rating starts with a Failure Modes, Effects and Diagnostic Analysis (FMEDA), quantifying susceptibility of the design to transient faults (triggered perhaps by ionizing radiation), as mitigated by error detection or correction techniques such as ECC, triple modular redundancy and lock-step computing.

The analysis workhorse behind FMEDA is fault simulation, modeling transient faults as stuck-at faults in traditional design for test analysis. These engines are very capable but cannot exhaustively test every possible fault in a large circuit. However not all possible faults are detectable at the outputs of a function block. Formal analysis can make a significant contribution here by pre-filtering unobservable or otherwise unimportant faults before fault simulation starts.

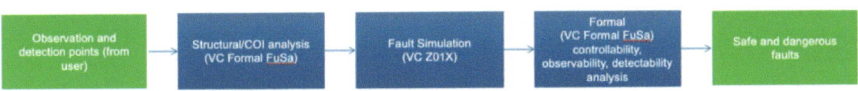

Using formal to filter out fault that need not be verified in safety checks

This filtering is critical to high-quality safety signoff because the total set of all possible faults in a circuit is far too large to analyze in fault simulation. Reducing this set to only important faults through formal methods, separating safe faults (which have no real impact) from dangerous faults (which could have serious impact), makes the problem tractable. Ensuring that SoC integrators can confidently certify ASIL readiness at the appropriate level.

Chapter 6 Apps Requiring More Hands-On Experience

The apps discussed in the previous chapter range in complexity from simple to use, to relatively modest complexity. The applications we discuss in this chapter aim to simplify formal analysis of more complex problems through automation, but generally require deeper understanding of the application space and methods to work through proof strategies. Nevertheless, these applications are much easier to use than using unassisted property verification to meet a similar goal.

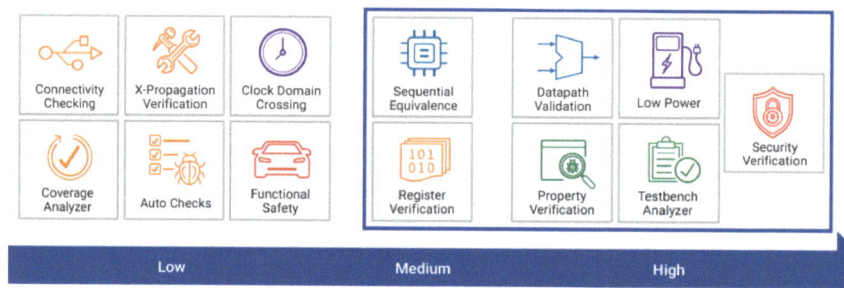

Apps ordered by ease of adoption
In this chapter we review the higher effort apps on the right

Sequential equivalence

What is sequential equivalence checking, why is it important and why isn't it covered in standard equivalence checking? We talked about equivalence checking in chapter 2. That type of check is used to verify that the gate-level model generated though synthesis is functionally equivalent to the source RTL (that is, that synthesis didn't break the functionality). Because of the way synthesis works, this only needs to check equivalence of combinational logic between registers[57].

However, some design changes require sequential modifications. One such case involves adding clock gating for power optimization. The ultimate functionality should remain the same in the enabled state, but cycle shifts (latency) may be added when enabling and disabling the clock since new register stages have been introduced. What must be compared between the "before" and "after" logic now includes logic which would confuse conventional equivalence checking[58].

Clock gating optimization – before gating (above) and after (below)

---

[57] This isn't strictly true. Synthesis engines may re-time logic across registers, so equivalence checkers must understand these cases also. But allowed deviation from the basic principle is limited.

[58] The gated and ungated functionalities are logically different but in a well-controlled way

Since sequential equivalence checking is about comparing two designs, you need a source RTL (before you make changes) and an implementation RTL (with clock gating changes). Just as when you compare two versions of the same Word document, you'll generally see these two RTLs side by side in the debugger window. The assertion generation part of SEQ is still hands-free – the app will generate these for you.

The debug part of SEQ (looking at results) is based on Synopsys Verdi and is going to look very familiar to anyone who uses that tool for simulation debug. You look at detected problems which leads you to waveforms, then you'll cross-probe to RTL and trace back to root-causes. The only aspects that are a little different here are that you are looking at two sets of data (the specification/original version and the implemented version) and there are some extra debug features in Synopsys Verdi, such as sequential trace-back to a root cause. But all of this will still be very familiar to a verification engineer.

The only part of the task that gets more (formally) hands-on is the proving phase; you can think of this as a first introduction to the details of formal proving. In conventional equivalence checking, comparisons are quite bounded – tracing most typically stops at sequential elements like registers. But in sequential equivalence problems, as in clock-gating verification, tracing may need to go through multiple sequential elements. This makes the potential complexity of the problem very large, which in turn can result in proofs which do not complete in reasonable time or memory. Proofs terminations of this type are our earlier-mentioned *inconclusives* – no counter examples have been found as far as the proof was able to reach, but equivalence has not been conclusively proven.

Sequential analysis (SEQ) provides a nicely automated way to overcome this by automatically decomposing challenging problems into smaller sub-problems, then proving equivalence between these sub-problems. You can track this progress in Synopsys Verdi. Where sub-problems converge, those sub-proofs are complete. Where they fail to converge or where an apparent mismatch (assertion fail) is found, that signals need for a new (and automatically triggered) decomposition. The app will continue to try to find new decompositions, so this part continues to be largely hands-free (you can still watch progress/status in Synopsys Verdi). In the simpler cases, the sub-problems converge, leading to a proof (or counter example) at the full problem level.

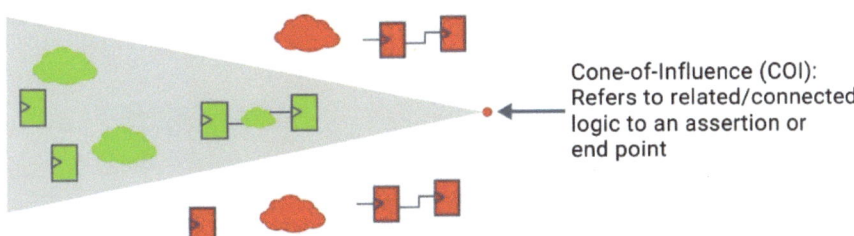

The cone of influence (COI) of a property
Light-colored logic is in the cone of influence, darker logic is not

Where you may have to get involved is if it becomes clear that certain parts of the design are resisting convergence. This could be caused by memories or counters or other complex blocks in the **cone of influence** for a proof, since these typically lead to explosion in size in formal methods. At this point you may need to turn to the tools we mentioned earlier (abstraction, invariants, constraints). If you are planning to be a hands-on verification engineer, fear not; learning how to use these techniques doesn't require an advanced degree; this is just more skills development to add to your arsenal of verification expertise.

Register verification

Almost all CPU-based designs today are memory-mapped. Components in the design are controlled, written to and read through registers, potentially tens of thousands of registers. That register logic can be quite complex – registers which are read/write or read-only or are cleared on read, or many other possibilities. These registers come in different sizes, split into ranges (bitfields) with different functions, each with that wide variety of possible read/write properties. And there are more variations. Getting this right isn't just important to hardware verification. Software communicates with the hardware through these registers, so absolute correctness in behavior is vital. Register verification apps automate this checking against an IP-XACT or similar specification. A typical flow is shown below, in which register implementations can be checked against definitions in IP-XACT or RALF.

Register verification

Low power

Mobile and green technology demands, costs and reliability have all driven power to become a high priority in delivering competitive designs. This led to creation of the UPF standard for defining the power intent supporting power-aware implementation. We quickly discovered that these intent descriptions for SoC designs can be extremely complex, which in turn prompted new verification methodologies to validate the correctness of the described intent.

Some of this intent can be verified statically. When crossing from one voltage domain to another, power intent should require a level-shifter. Or an input to one block from another block in a power-switchable domain should have power intent specifying isolation logic. These needs are not use-case dependent so can be checked statically.

Formally verifying isolation controls based on power state

But think a little more about the isolation case – under what circumstances should isolation be enabled? This does depend on the use-case, so cannot be verified statically. Commonly verification teams further enhance their simulation/emulation testbenches to check these cases, which provides some level of increased confidence in coverage. But there's an obvious limitation in that approach which goes back to the reason formal methods first became popular. Creating, running, and debugging a huge number of test cases to verify the correct behavior of the design *without power considerations* is already a huge task. Now imagine having to repeat that analysis across each of the possible power configurations for a design. Talk about combinational explosion - this would be completely impractical.

Formal verification apps can help here. These look at power-state switching expressions to determine if there may be conflicts, providing you with confidence that, at least to this extent, all possible behaviors have been proven correct.

One last but very important point. Many of the checks performed by Apps in this and the previous chapters must continue to be valid even in the presence of power switching. Connectivity checks, functional lint checks, sequential equivalence checks, even custom formal property checks must continue to remain valid (as appropriate) across different power states. The ability to make these checks while also understanding power intent is a critical contribution from formal analysis, since dynamic checking generally cannot attempt this level of coverage.

Security verification

Security is a moving target; threats continue to evolve, therefore defenses must also evolve. Unfortunately, security is an area where even 99% coverage isn't good enough. Hacks rely especially on rare and obscure weaknesses, so verification must be as close to perfect as possible. Which means that formal methods are the only acceptable way to signoff.

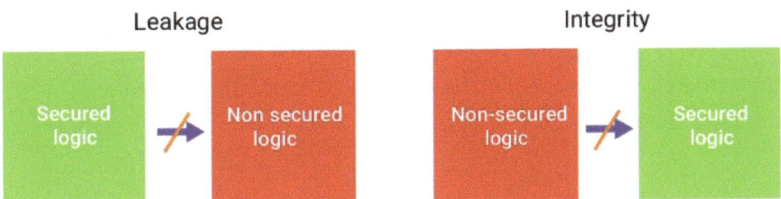

Checking for data leakage or data integrity compromise between secure and non-secure domains

The scope of what might possibly need to be checked can be large, including defenses against embedded hardware Trojans[59] and side-channel attacks[60]. However, the great majority of commercial activity is centered on attacks where software in an insecure domain tries to get access to data in a secure domain, by exploiting design or architecture weaknesses. As we have seen recently with the Meltdown and Spectre bugs[61], hardware is not immune to security problems, despite all the advanced techniques that are already being used to limit attacks.

Given the range of security techniques that can be applied, and the intentionally limited scope of secure areas in many cases (to reduce the attack surface), this domain is a natural for app support. However, the nature of these checks must change in a couple of important ways.

First, we want to make as few assumptions as possible about how data might be leaked out of or injected into a secure domain through any reasonably bounded sequence of operations. These are general information flow checks, looking for potentially leaky

---

[59] https://en.wikipedia.org/wiki/Hardware_Trojan
[60] https://en.wikipedia.org/wiki/Side-channel_attack
[61] https://googleprojectzero.blogspot.com/

paths.

A second consideration is compliance with emerging standards such as the Mitre Common Weaknesses Enumeration (CWE) and the Accellera SA-EDI standard. Industry expectations are starting to lean more on these definitions as a minimum requirement, which will drive signoff requirements.

Testbench analyzer

Kiran mentioned the concept of formal signoff in the foreword to this book. Meeting that objective requires not only a good set of assertions but also proof that the formal testbench is compete and not over-constrained. The Synopsys VC Formal FTA app checks for missing or incorrect properties and constraints.

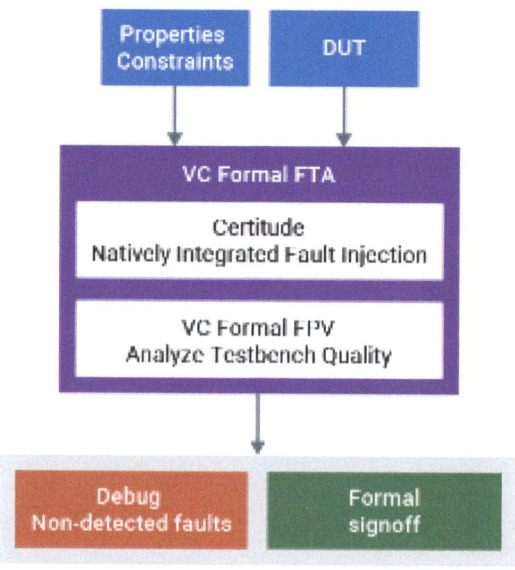

Formal testbench analysis

Meeting a signoff objective requires not only checking a testbench as defined, but also verifying its coverage completeness in a conceptually similar sense to coverage completeness in dynamic verification. The FTA app accomplishes this through mutation analysis. This uses the Synopsys Certitude technology to inject design errors into the RTL. A high coverage testbench should detect these errors. Failing to detect them may indicate insufficient assertions, or over-constraints in the design. Using FTYA analysis increases confidence in testbench quality for formal signoff.

Chapter 7 Formal Methods for Datapath Elements

The exploding importance of custom datapath functions

What are datapath functions[62]? Think of the buttons available on a calculator, particularly a scientific calculator. You'd expect to see add, subtract, multiply and divide, also square root, exponential, and log functions. Probably also trig functions like sin and hyperbolic functions like tanh.

A scientific calculator

When computing was primarily delivered by big processors, such functions were handled either within the processor or in a closely coupled co-processor.

Datapath functions also appear in graphics processors (GPUs) and digital signal processors (DSPs). GPUs must perform a lot of math-intensive operations: video encoding and decoding, 3D rendering, rotation and shading calculations are just a few

---

[62] More precisely, in this context these are datapath "manipulation/transformation" functions such as math functions. Datapath "transportation" functions can be handled in normal formal property verification.

examples. DSPs are often used on digitized analog signals to provide complex and programmable signal transformation functions: filtering, Fourier transformations and matrix operations are common examples here. Beyond these specialized functions, any other device requiring math calculations would delegate the requirement to one of these devices.

But not anymore. Boundless demand for intelligence, connectivity and enhanced user experience everywhere is pushing richer computation into mobile and remote devices, in part driven by demand for faster response times at low power and low cost. AI is a critical component in many of these devices, using accelerators built on multiply-accumulate (accumulate here means add) functions, known as MACs. For training applications in datacenters MACs could be operating on 32-bit floating point numbers. In inference in edge devices, they may operate on 4-bit numbers.

Widths can vary widely and are not limited to powers of 2. Number types can range over integer, different flavors of float, and fixed point. Number types are assigned and "quantized" to whatever best meets application-specific tradeoffs of power, performance, and area. Different choices may be used in different layers. Compute engines are often organized in systolic arrays to maximize throughput, with math computation embedded in each processing element in the array. The activation stage in neural nets requires functions like tanh (hyperbolic tangent), or sigmoid functions. You get the idea - AI demands a lot of embedded special purpose math engines. One-size-fits-all math functions have no place in this domain!

Simultaneous localization and mapping (SLAM) used to be a niche function in robotics, but now robots are everywhere. In home robotic vacuums, delivering medicine and food in hospitals, guiding drones, even guiding self-parking for cars. SLAM requires a huge amount of linear equation solving (among other calculations), which depends on very intensive matrix math, in fixed point or floating point.

Pedestrian, road sign and lane detection are commonplace in modern cars, driven by image-based AI. Collision detection and enhanced guidance in poor weather rely on radar or other techniques for which detection uses inferencing on 4-D point clouds, adding inferred velocities from Doppler shifts to 3D positioning. More math is needed here and must also deliver fast compute times for safety-critical responses.

5G call quality is enhanced by a feature called beam forming. Cell phones and base-stations collaborate to determine which station can provide the best connection, then optimize between the two by effectively "zooming" to the best direction through constructive interference between signals received or transmitted at antenna arrays. A similar concept can be used for audio beam forming, between a smart speaker and a person giving commands to that device. Both applications need fast math computation to detect and direct the direction of transmission and/or reception.

There are many more examples, in 3D audio, in VR/AR and others. All depend on fast embedded math computation within a wide range of number types and widths, many needing careful tradeoffs between performance, power, and cost. Custom datapath elements are now everywhere and they are driving a corresponding interest in tools to verify datapaths. As one indicator of market demand, at Synopsys we're finding verification customers first wanting to buy simulators, next coming back for emulators, and then coming back for datapath verification tools!

Verifying datapath elements is a challenge

Remember the section earlier in the book on the verification treadmill and the price of not being perfect? Intel found a floating-point bug in a device in production, an issue that cost them both dollars and a dent in their reputation because they missed a problem in datapath design. Simulation in theory could find such a problem, but it isn't complete. For math operations, "right almost all the time" isn't good enough; they must correctly work every time.

We now know the answer to this issue – use formal, right? But there is another problem. Standard formal methods are generally not effective at verifying datapath functions. It's worth digging a little deeper into this topic to understand why, because that will highlight a need for a different approach – the subject of the next section in this chapter.

There are multiple ways of implementing math functions in hardware, but all face an unavoidable reality. We like precision in our math, using 32-bit numbers at minimum, more commonly now 64-bit numbers. Proving that a 64-bit by 64-bit integer multiplier works correctly requires proving each of 2^{128} or 3×10^{38} cases. In formal methods, we use symbolic methods, but even so the intrinsic complexity of the problem is overwhelming.

The complexity of an integer multiply grows rapidly with word size

If formal methods are to have value here, they must provide complete proofs. Not being able to meet this tough goal had turned most verification teams back to simulation for datapath testplan objectives, leaning heavily on as much creativity as they can muster to search out corner cases. While still knowing that no matter how clever they might be, errors can slip through.

Back to equivalence checking – with a twist

Look again at the question posed earlier in this book, "What do we mean by correctness?" Property checking – verifying a set of behaviors you expect to hold under all circumstances – is one way to answer this question, but it is not the only way. A different and more complete approach is to check equivalence with a trusted specification. However, trusted specifications are often hard to find. In most cases, this limitation has restricted equivalence checking to comparison between verified (and therefore trusted) RTL and gate level implementations.

Luckily, math operations have independent trusted specifications. For example, the standard IEEE 754 precisely defines expectations for floating point operations and academia provided a Soft-float library of standard C-models. In this case, an RTL implementation of an operator can be checked for equivalence against the corresponding C reference implementation.

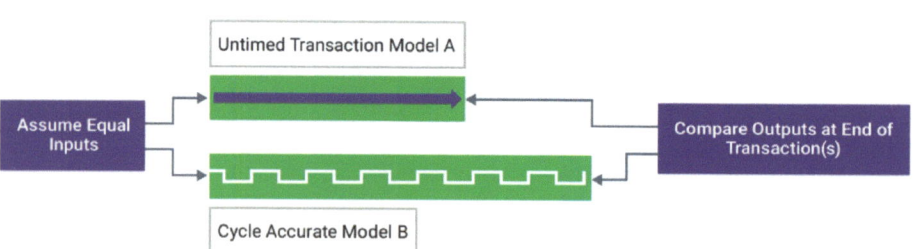
Establishing transaction equivalence

But the C reference is neither timed nor guaranteed to correspond at a detailed level with a hardware implementation; equivalence checking is this context must therefore be at a transaction level. Proving that, independent of internal algorithms, the specification and implementation will generate the same outputs given the same inputs. We'll use Synopsys VC Formal DPV as our reference in what follows to avoid confusing generalizations. The general approach to this method is conceptually like RTL/gate equivalence checking, except that in this instance the reference design is in C (or C++) and no attention is paid to differences in when the output is generated.

Formal methods appeared earlier in software verification than in hardware, so the infrastructure for proving in each domain is already well developed. Innovations here are in connecting the hardware and software model representations and in providing industrial-level support for real-world proof management. We'll touch on some of these aspects in the next section.

It's worth asking why equivalence methods work in general where property checking doesn't. Why don't they run into similar size problems? The success of equivalence methods in RTL/gate checking on large circuits provides a hint. These are effective because the large objective can be decomposed into many smaller and therefore very tractable sub-proofs. Math functions have a quite different structure, but a similar principle can still be applied, dividing the large problem into smaller sub-objectives. With suitable cross-checks, those sub-proofs can then be connected into a proof for the full operation, making proving the correctness of a 64-bit floating point multiplier completely feasible. Think about that for a second. A ***complete, signoff*** proof of the correctness of a large production multiplier, against a definitive specification. No wonder these methods are attracting attention!

That said, using decomposition techniques correctly requires experience in knowing where and how to partition. It also requires care in verifying assumptions across partition boundaries. These are not simple choices. A production-ready prover should provide significant help and automation to support this activity and prover products should be measured not only on correctness but on how much they simplify these complex choices. A very helpful fact is that the proof domain spans a limited set of well-defined algorithms, which should allow for a high-level of assistance.

Practical considerations in datapath verification

Knowing what you now know about practical application of formal methods, you won't be surprised to learn that datapath verification isn't completely hands-free. However, the methods used here should be quite familiar from our earlier discussions on property verification. We'll use a floating-point multiplier to illustrate.

The basic algorithm is quite simple: add the exponents, multiply the mantissas, do a little adjustment, for rounding and normalization for example, and you have the result. Each of these operations is simpler than the total algorithm so each can be handled as a sub-proof. For example, the mantissa multiplication is an integer multiplication.

A floating point multiply implementation

A C-algorithm for a floating point multiply should follow a similar structure, allowing an intelligent prover to automate finding correspondence points between the C definition and the RTL implementation. Sub-dividing the larger proof objective into smaller proofs with already known strategies. For example, the mantissa multiply is an integer multiply. If the implementation can be guaranteed valid, then that fact can be used to simplify proof at the next level. This is the *assume/guarantee* approach we saw earlier. The step to prove validity of the integer multiply can leverage another technique we have already seen – *case splitting*. For example, the integer multiply can be case analyzed on the top 4 bits (16 cases), proving each simpler case in turn. Proof across all 16 cases therefore proves the full multiplier function.

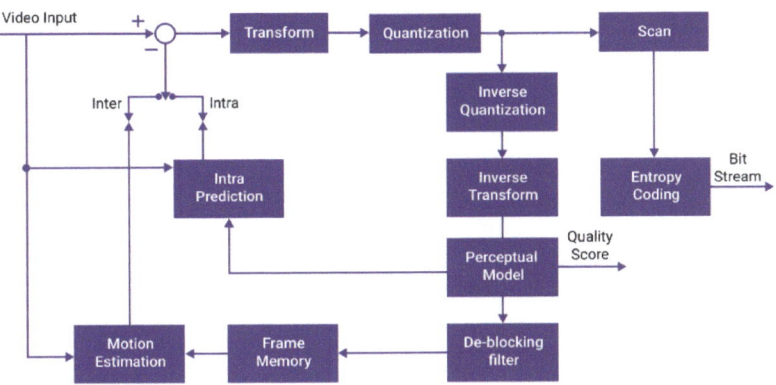

A video encoder implementation

Why not simply run these proofs on IP library elements, then design only with proven elements? Generators for most IPs will be highly configurable, for word widths and multiple other options. Pre-validating all possible configurations is totally impractical, and validation post-configuration becomes essential. Further, consider the complexity of verifying a subsystem like a video encoder built around multiple different datapath functions, finely tuned in configuration to support competitive performance, power, and cost goals[63]. Manually setting up proofs for each of these functions, per

---

[63] A similar approach would apply to almost any function in an ISP pipeline, such as de-mosaicing and dynamic range management.

configuration change, would be very painful. Intelligent automation is a critical technology needed to accelerate those component proofs in support of validating the complete function.

Recap

These apps, from formal linting to datapath verification, simplify verification for a wide range of tasks, but they don't represent all that can be accomplished with formal verification. Use for more complex or more unique problems requires you to write your own assertions and may require some level of involvement in getting to proof-convergence through abstractions, constraints and other methods. We'll dip into this topic later in the book.

Also, while we've said this before, we'll repeat it again. A characteristic of the bounded model checking method at the core of most formal methods is that while a proof (an assertion passes or fails) or a counter example will often be found within the bound of proving, this is not guaranteed. We started to see this in our short introduction to the SEQ app. In these cases, to get to a proof, you can increase the bound, or do more abstraction, or add more constraints, or any combination of these. Or you can ask whether the depth to which you have checked, with the abstractions and constraints you already have, is sufficient to declare the proof acceptable. We'll look a little more at this topic later in the book.

Now we're caught up with the basics, it's time to look at how your peers in the semiconductor industry are using formal verification today.

Or in GPUs for functions such as shading or de-interlacing. Or in AI engines for processing elements in an array, for weight compression/decompression and for matrix operations. All built around math-centric algorithms.

Chapter 8 The Role of Formal in Design Today

Adoption

No matter how clever or easy to use a technology may be, the only measure of success that matters to both vendors and users is real-world effectiveness, as indicated by adoption and growth; how many people / organizations are using it and how quickly is that usage spreading? Multiple surveys in the industry indicate that both are robust and have moved beyond early expert adoption.

For example, one survey[64] shows almost 35% of reporting projects using formal apps in verification and over 40% using custom property-checking. Among these, app-based verification contributed to significant growth in formal usage from 2012 onward, followed by accelerating pickup in custom checks as indicated in 2020 surveys, showing perhaps that growing familiarity with app-based approaches is making verification teams more comfortable in moving also to those custom applications.

Formal verification is accelerating even in FPGA-based design[65], at or over 20% of projects in 2020, no doubt reflecting the increasingly complex nature of FPGA SoCs and the need for development teams to check *cannot-fail* assertions as completely as in ASIC designs (especially in mil-aero applications where FPGAs are widely used, and now even in ADAS, where safety and reliability expectations are also very high).

Regarding business growth, informal feedback supports a view that the top 20 companies (by revenue) designing complex SoCs are already investing $80-100M/year in their formal verification flows. Perhaps twice that much is being invested across the entire user-base. Also noteworthy is the range of applications; designers are

---

[64] Verification Trends 2020
[65] 2020 Wilson Research Group Functional Verification Study

applying formal methods across the spectrum, from processors and graphics, to wireless, networking and storage, image processing and recognition. Indeed, there is no obvious reason not to use formal in most areas of digital design today.

Among organizations that have established teams dedicated to formal-proving, it is common to hear that these methods may address 40% or more of the total verification burden[66]. Formal is no longer a niche technology – it is now carrying a verification load similar to other verification methods.

An important question is *why* these teams are using formal methods. A cynic might hold that "it's hot so we better do some to keep management off our backs". We doubt this accounts for any significant usage; most verification teams are under such intense pressure they don't have time for "show" projects. A much more common reason heard from verification leads is growing concern that as design complexity continues to advance, they already see or anticipate important verification problems moving beyond the reach of simulation-based methods. They are allocating budget, resource, and schedule to ramping up in formal because they have no choice. Rare or intermittent problems which escape to the field are often out of reach of "just try harder in simulation" but can spiral into disproportionate costs and reputational damage. Design and verification teams are increasingly turning to formal methods[67] [68] [69] [70] as one way to shake out those difficult problems.

---

[66] See the foreword, also Simulation and Formal – Finding the Right Balance
[67] Adoption, Architecture and Origami
[68] System-Level Formal
[69] DVCon 2017 Making Formal Property Verification Mainstream: An Intel® Graphics Experience
[70] DVCon 2018 Architectural Formal Verification of System-Level Deadlocks: Qualcomm and Oski

Size constraints

Whenever formal methods are discussed, you will hear some dismiss them as "only for small problems". The exhaustive nature of formal proving certainly limits the size of the state-space that can be considered in a proof, but this should be offset against the potential for abstraction, automated in many high-value apps, and of course continuing improvements in proving technologies, especially around advances in heuristics.

Given that, what are realistic size-limits? It depends. A full SoC state-space far exceeds the practical limits of any current formal method, but there are multiple useful SoC-level problems which are routinely tackled, especially in apps like connectivity checking where block abstraction is easily automated. Cache-coherency verification is an example of a system or large-subsystem-level problem also commonly reduced to a manageable level through abstraction.

Unquestionably custom property-checking is most commonly used at the IP level, since pre-tested constraints and heuristics may not be available and therefore a limited state space makes proving less challenging. Even at this level, abstraction is often needed to simplify memories and datapath elements. However, this need is really a feature of the formal approach rather than a limitation. You do this to enable formal proving on other parts of the logic, remembering that simulation may also be limited in exhaustively proving through such cases and often depends on abstractions such as behavioral models.

Granting all these points, what is the real story on state-space capacity in formal tools? This is a moving target and you'll no doubt hear different ways to calculate from other sources, but one approach we like for its simplicity is this. First, size limits are a function of the property being tested. The size of the cone-of-influence (COI) of that property is much more important than the design size.

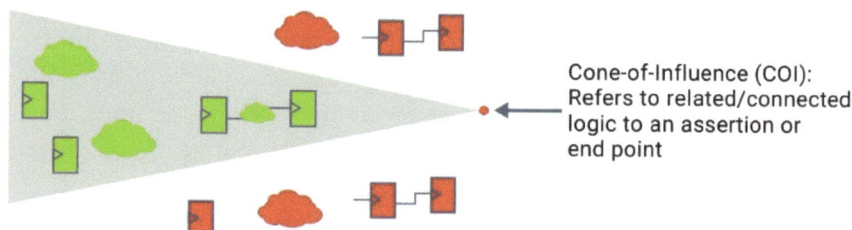

Reminder: cone of influence for a property

Given this, the problem size of the proof you are facing is determined by the number of inputs and state elements in that COI. Formal tools will often calculate these values for you. Within these bounds, one experienced user in the industry[71] cites **40K state elements** as a practical limit as of 2017.

Popular targets for property checking

Clearly, most teams start and continue with apps. These add obvious value in covering important parts of the verification plan (connectivity, coverage, registers, etc.), they're an easier place to start than custom property-checking and they help you incrementally build expertise in using formal tools. Adding custom property checks comes later and is commonly viewed as an incremental extension to app-based checking to handle specialized and proprietary objectives.

Good targets

When looking at block-centric checks, whether app-based or custom, some objectives fit well with formal while others are

---

[71] Erik Seligman in Formal Verification: An Essential Toolkit for Modern VLSI Design

better handled in simulation. Now we know that formal works well with control logic and with datapath logic, let's look at each in a bit more detail.

First, formal works well with control-intensive blocks which have lots of states and transitions, with complex conditions for transitions and possibilities. Think of very complex FSMs or multiple inter-operating FSMs. Common blocks in this class include arbiters, controllers of various types (memory controllers, power management controller, traffic controller, interrupt controller, etc.) and schedulers (round robin, fair etc.) for quality of service.

A representative SoC – arrows indicate some areas formal will likely find a role

Similarly, in networks, verification of bus bridge behavior (from master to bus, from slave to bus or from one bus protocol to another) is a good candidate, ensuring that data isn't lost due to FIFO overflow in the bridge, or loss of request or acknowledge signals from/to the bridge. Equally proving correct operation of network traffic management techniques, such as credit management, are excellent candidates for formal proving. Cache coherent networks are a very hot topic in modern multi-processor systems and pretty much depend on a significant level of formal validation.

Data transport systems are good candidates, where data is being moved from one place to another rather than being transformed in computation. Examples can be found in crossbar switches and networks-on-chip (NoCs).

An area of verification that is always of high interest for formal methods is interface checking, especially for protocol validation. This usage should be approached with some caution. Verifying simple interfaces is not too onerous; for example, coding assertions and constraints for the AMBA APB protocol can be completed in a matter of hours. However, checking for more sophisticated interfaces can become very complex very quickly; coding a correct and complete set of assertions and constraints for AXI3 (or higher) could take weeks.

The best approach for interface-verification is to use pre-packaged assertion IP whenever possible; these are already setup with assertions and constraints proven to work effectively in a wide variety of contexts. Supporting documentation will tell you what aspects of operation are covered and what (if any) you need to handle separately in simulation-based testing.

We went into some depth in the last chapter on formal verification for datapath functions. These features are now very much production proven and include capabilities across a wide range of functions. In a similar vein, in the first edition we advised against using formal on more complex functions like image codecs which depend heavily on datapath functions. Complex blocks like this are now very much in range of formal methods when suitably decomposed.

Targets to avoid

Certain tests on interfaces which require proving over very long sequences (such as for PCIe, MIPI, SATA or HDMI) are not practical. (Formal checks on the non-sequential portion of those designs are still useful.) Standard formal methods depend on proving/disproving within relatively limited clock-cycle bounds; if a proof depends on analysis over sequences of many thousands of clock-cycles, the state-space can become unmanageable. Even here there are already bug-hunting methodologies for directed searching across many cycles.

Also note that formal methods are naturally designed to provide binary (pass/fail) responses rather than statistical responses. They are not well-suited to quality-of-service (QoS) verification, except at the fringes. If you want to know whether your system can ever deadlock, or will acknowledge a request within some time bound, formal can help. But if you want to know whether each requestor is handled fairly across different traffic profiles, that is a problem for simulation and statistical analysis, not a problem for formal.

Product and service solution providers

Apologies to any solution providers we may have missed; we believe this list is fairly exhaustive.

Synopsys (VC Formal)
>Synopsys built its formal product lines entirely in-house, originally represented by their Magellan product, later adding the Hector high-level equivalence-checking tool. It subsequently developed Synopsys VC Formal as a completely new product, which now has largely superseded Magellan. Synopsys has a broad product and solutions offering.

Cadence (Jasper Gold)
Cadence started in this space with BlackTie from their Verplex acquisition, evolving into Incisive Formal Verifier. Later they acquired Jasper Design Automation. Jasper is generally credited with raising the profile of formal verification, in part through intensive customer support/services model and in part by popularizing the app concept. Cadence has a broad product and solutions offering.

Siemens-EDA (Questa Formal)
Siemens-EDA (originally Mentor) became prominent in formal through their acquisition of 0-In, which made them the leading player in formal in the early days. They added further formal capability in sequential equivalence checking and C to RTL equivalence checking through initial engagement with Calypto and later acquisition. Over time they have rebranded this solution as Questa Formal. Siemens-EDA has a broad product and solutions offering.

OneSpin (360 MV family)
OneSpin was (appropriately) spun out of Infineon from an internal formal verification team. OneSpin has a broad product and solutions offering and has been acquired by Siemens-EDA.

Others
Real Intent started with a strong focus in formal verification. Market presence is now primarily around their static verification product, while formal capabilities are centered in the Ascent family.

IBM was an early leader in Formal Verification with its own in-house tools RuleBase and SixthSense, which are still being used in limited fashion at IBM

Chapter 9 Adding Formal to Your Flow

This section is written primarily for verification managers and verification team leaders, to guide managing formal verification in your organization, ***without having to be a formal expert***. We're not going to spend any time on how to setup and run tools; we are going to spend a little time on how to plan for formal in the larger testplan, how to understand and guide formal team members and how to measure progress towards a signoff goal. That said, even though this isn't about tools, methodologies may vary between vendors. Again, we'll be guided by the Synopsys *VC Formal* flow. Many of these concepts will carry over in some manner to other flows but you should check with your vendor for possible differences.

Organic skills growth

Despite the obvious advantages in adding formal methodologies to a verification flow, simply jumping in as you might with any other new capability (buy the tool, train on the tool, start to apply on live designs) doesn't always work out very well, in part because formal requires a mental switch from familiar simulation methods, also because it may be viewed with suspicion by the rest of the verification team. On the other hand, postponing adoption until you can hire formal experts (who we'll call black-belts) generally becomes a decision not to make a decision, and that can be an expensive mistake if you already anticipate hard problems overwhelming simulation-based flows.

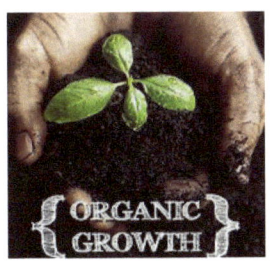

Formal black-belts are still rare today and have little difficulty finding jobs with big-name companies. If you want to build formal capability in your team, you almost certainly need to grow it internally and/or recruit whatever brown-belts you can find. Experiences learned from companies who have done this successfully suggest some common steps towards adoption:

- *Encouragement*: Verification team members are encouraged to learn more about formal verification successes, applications and case studies through conferences, EDA vendor meetings and personal learning. Make a little white space in those crushing product schedules.

- *Find and grow a champion*: When one of those people shows an interest in championing formal methods, encourage, and support them with a small team, budget, schedule and supervision for planning, discussion and feedback. This shouldn't be difficult - most engineers want to learn new skills. Most important, even more than background in formal, is willingness to take risks.

- *Freedom to learn and develop a methodology*: Managers should provide plenty of flexibility for learning and finding a path to demonstrate initial successes. No-one gets hung up on "we already showed that in simulation" or "we got there much faster using simulation". The goal is to learn and develop a working flow, especially for the apps. Experienced verification teams have been quite open about this phase taking a year.

- *Pick the best targets*: We'll say it again – start with the Apps and AIPs. These are by far the easiest path to develop expertise and show value. Over time you can graduate to adding custom property-checking, but there's no need to rush to that point.

- ***Develop metrics***: It is always important to develop metrics for what is being proven, but early flexibility must be allowed in these metrics – teams need to grow into what they can prove and should be allowed to start with simple metrics like "no unresolved counter examples". Capture data on effort (engineering, CPU hours, etc.) to help develop ROI cases. Over time, push for harder limits, e.g., around coverage.

- ***Hands-on big picture supervision***: An important point here is that the verification supervisor (you perhaps) should not be completely hands-off. You don't necessarily have to run the tools or understand the details, but you do need to keep connecting metrics and success back to the big picture – is this ultimately headed towards production usefulness? At the same time, you can become more familiar with the concepts of the domain, so you can learn (at a high-level) how to question approaches and provide guidance.

- ***Socializing in the design and verification teams***: It is also very important to socialize progress within the larger verification team and to gather constructive feedback. Fostering a collaborative environment with the formal team ensures everyone understands goals and benefits and helps the formal folks optimize their focus. You definitely don't want mainstream verifiers looking at this as low-value playtime. There is also real benefit in including designers in these discussions, first because formal verifiers may need their help in handling inconclusive proofs. If designers are not bought in, they will not provide adequate guidance to get to formal signoff. Second, there is value in promoting use of formal in RTL design; the formal team can help designers use the tools to explore behaviors in their RTL, ultimately helping them to hand-off higher quality IP to verification.

- ***Socializing in the management chain***: Successful adoption efforts have also been careful to socialize progress and goals further up the management chain. You might be surprised – even CEOs can be interested in what you are doing in formal, especially where it helps add to their product quality, safety, or security pitches.

- ***Review, refine, advance***: As in any good engineering project management, the first round should be followed by careful review and analysis: what worked well, what didn't, what should be attempted next and how metrics should be tightened to more effective levels. This review should definitely involve the larger verification team; the expectation should be that in the next round the formal team will contribute materially to verification closure. That said, be patient. Getting to productivity may take a few false starts. Once you start getting successes, that success will be contagious.

First-cut targets for formal verification

The essential components of your verification process don't change when you add formal methods to the mix. You build a verification plan, partitioned by blocks and behaviors to be verified. Within that plan, you will segregate the verification/coverage plan by appropriate verification technologies:
- Goals that should be a good fit for formal
- Goals that will work well with constrained-random (IPs, small subsystems)
- Goals that will work well with emulation (regressions, SW/HW co-verification)
- Goals that will work well with FPGA prototyping (regressions, SW/HW co-verification)
- Goals that must use simulation for other reasons (e.g. AMS verification)

When looking at candidates for formal, consider these cases:
- Any verification task for which an app already exists, like top-level connectivity checking, register checking and sequential equivalence checks around clock gating. This is the easiest place to start; it should be much easier than simulation and it will be more complete. These cases should be no-brainers.

- Verification plan line-items for which assertion IP are already available, for example the section(s) in the verification planning covering protocol compliance on interfaces such as the AMBA interfaces (ACE, AXI, AHB, etc.) and common IO standards (USB, aspects of PCIe, etc.)

- Any case where you know control complexity is very high and difficult to cover solely in dynamic testing (interdependent FSMs are generally good candidates)

- Blocks which have historically been problem-prone from release to release, exhibiting intermittent hangs, deadlocks or other issues

- Once your team has developed experience in the basics, think seriously about datapath functions. This is an area where formal can definitively beat dynamic techniques and convince even the most hardened sceptics.

And of course, remember good targets may be more about the test than the block. Any given block might be best served by leveraging multiple verification platforms, each targeting different tests. The "challenging blocks" case is a good example. Maybe these could benefit significantly from some carefully crafted formal proving in addition to dynamic testing.

Detailed planning

Once targets are identified, then you start planning detailed tests. While there is a concept of testbenches in formal verification, this term is primarily associated with the end-to-end verification concept we mentioned earlier in the book. Since this is a fairly advanced usage, you probably won't consider it in early stages of formal adoption. In app-based verification and your initial forays into custom property verification, this testbench concept is not so obviously relevant so we'll stick to calling these tests.

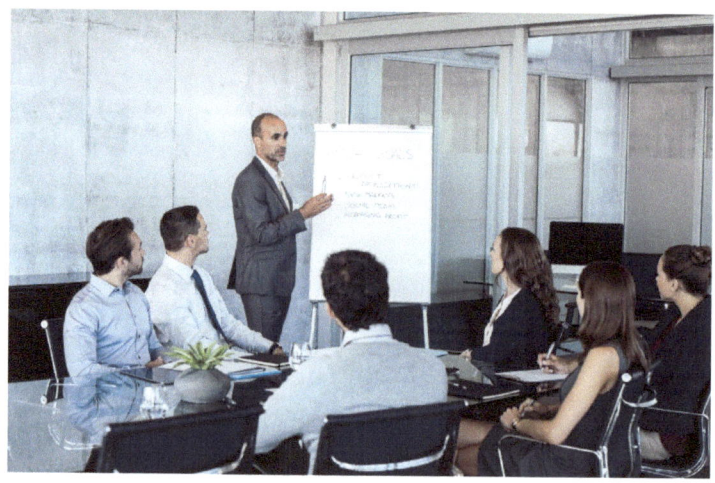

Just as in detailed planning for simulation, you should have a planning stage per block (viewing top also as a block). In each case, you want to start with plain English descriptions of checks you want to perform, for example "verify this state machine always returns to IDLE after interrupt". Get these descriptions right first, because they'll be easier to understand and debate than the SVA translations they will eventually become.

Once detailed planning for a test is complete, your formal experts can start building an executable test. This won't look much like the UVM bundles you know (and maybe love), running to thousands of lines of code. You should expect a few Tcl files, SVA files and bind files, in total running to a few hundreds of lines (lines of code similar to targeted simulation setups). Again, you can get to testbench concepts in formal verification, but that is definitely for more advanced users.

Monitoring progress

In simulation you ask team members to provide regular updates on how many bugs they found and what coverage they have reached. Metrics aren't so very different in formal verification. The team will find counter examples (CEXs/bugs), some in the design and some in their tests. They'll find these quickly at first, then more slowly as bugs in the design and in tests are shaken out – this should sound very familiar.

A nice surprise for verification managers is that formal verification will start finding bugs very quickly; you don't have the typical simulation ramp-up phase of getting the testbench working. You may also discover that formal will uncover bugs in lots of unexpected places. While simulation testing works deliberately through a plan, formal races out to the fringes and can expose bugs before simulation test, a very real plus for accelerating verification closure.

In other cases, the formal team will report completed proofs for some properties and inconclusive results for others. This is where progress monitoring diverges from simulation, getting into questions of whether a completed proof is valid (related to constraints that have been set for that proof) and what steps can be (or need to be) taken next with inconclusive proofs.

The metrics you want to monitor, by verification plan line item, are:
- How many properties have been developed (asserts, assumes, covers)

- How many failures and covers have been found, and trends on these
- How many completed proofs and trend (but first read the next section)
- How many inconclusive proofs and trend (but first read the section on bounded proofs)

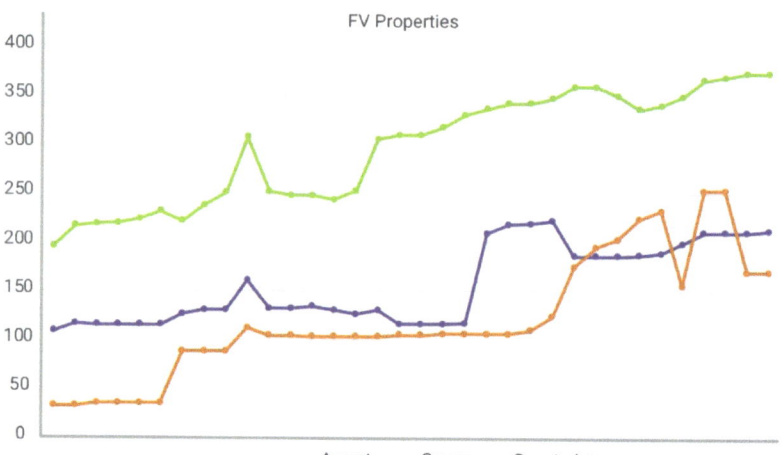

Tracking progress on developed assertion, constraint and cover properties

Tracking progress on assertion proof-status

Under-constraining and over-constraining

There's a Goldilocks aspect to constraints. You don't want too few, you don't want too many, you want just the right constraints. Under-constraining can cause formal engines to explore parts of the state space which are not meaningful for any practical use of the DUT (or perhaps for the usage you intend). That can have bad consequences. Analysis may report spurious counter examples (CEXs), which are failures in ways that are unrealistic. Or analysis may simply fail, running out of time or memory. Either way this takes engineer and machine time to analyze and to debug while not really advancing coverage.

The smallest area is an over-constrained state space, missing some bugs
The next larger area is the ideal –
a reachable and legal space

Good understanding of the modes of operation of the DUT or protocols or handshaking used by the blocks is the best way to avoid this problem. If you're finding bogus CEXs or a run won't converge, perhaps you need to add constraints – after discussion with the designer or architect. Where appropriate, one suggestion is to use pre-built Assertion IP (AIP) which have built-in constraints. This will help avoid spurious failures and get you up and running more quickly.

On the other hand, you can over-constrain, creating conditions which prevent exercise of realistic behaviors. There are some legitimate reasons to over-constrain during setup or bug-hunting, but not in signoff. In the Synopsys VC Formal flow, you can check for this possibility by running over-constraint analysis for each property. You can also check using reachability analysis for cover properties you know you should be able to reach; if you can't reach them, an over-constraint may be to blame.

Bounded proofs (inconclusives)

As we have no doubt tediously repeated again and again, inconclusive results are a fact of life in formal verification. Remember that verification is hard; there are no silver bullet methods to automatically verify everything. Formal methods provide a way to exhaustively verify properties out to some defined clock-cycle depth (proof depth). Which means that it is possible for such a checker to hit that bound without finding a counter example and without completing a proof that the check passes.

In earlier times, inconclusives were considered a real barrier to formal adoption – if a result was inconclusive, surely that meant it wasn't useful? Over time a more constructive and actionable viewpoint emerged[72], starting with a view that these were **bounded proofs** rather than inconclusives. It's now much more **common to view a bounded proof as fully acceptable for signoff, as long as the bound is well justified**[73].

---

[72] See for example "Signoff with Bounded Formal Verification Proofs", best paper award in DVCon 2014
[73] Formal Verification: An Essential Toolkit for Modern VLSI Design

This is understandable. The tool has exhaustively proved the absence of a counter example out to the proof depth; that is already valuable information. It is very likely that at some increased proof depth, the proof would be conclusive. In some cases, it is quite possible that this proof at this depth is already sufficient or would certainly be sufficient if extended out to a slightly deeper search.

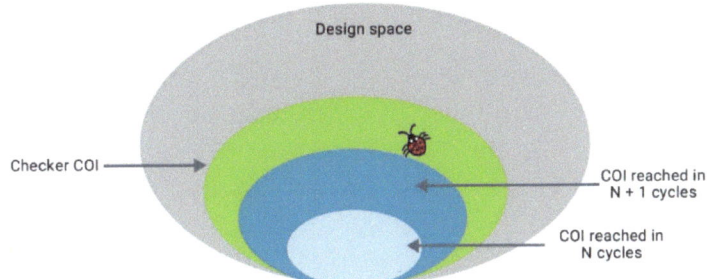

Exploring limits on a bounded proof

Of course, you can't just hope this is true; you need to know if the proof you have is practically sufficient or if you must work harder. A starting point is to look at a bounded coverage analysis, ideally also providing an incremental cover analysis. Coverage as a percentage of the property cone of influence is a good starting indicator. If this isn't high, look at the delta in coverage over the last two steps. A significant increase suggests that it may be worth trying again with an increased proof depth.

Another commonly used technique is to create cover properties for "interesting" corner cases. Finding what proof depth is required to hit these properties can be a good indicator of a minimum required depth[74].

---

[74] If you want to test your skill, you can try estimating the maximum number of cycles required to prove an assertion - with knowledgeable design help, of course. You may find in this exercise that you are missing an important constraint or two – which might further help in bringing proofs to closure.

Whenever you are exploring what depth may work, eventually you should have a discussion with the designer to agree on the "design/property radius" (the depth) they would consider acceptable. If the bound is not practical for formal proving, then you must consider abstractions to reduce the complexity of the problem space; that's the subject of the next section.

Manually guided proofs

It is quite possible that in some cases, an attempt to prove a property on a design will fail to complete because a full proof would exceed time or memory limits. We talked earlier about various methods to manage these cases – abstractions, decomposing a problem with assume-guarantee properties at interfaces between the sub-problems, and case-splitting with constraints.

You should expect that your team will need to resort to this kind of guidance in more than a few cases. But you'll be happy to hear that formal tools now go a long way towards simplifying this task. For example, tools can often auto-detect candidates for abstraction, such as datapath elements (though replacing logic with an inferred abstraction typically requires engineer approval). Decomposition and managing the assume-guarantee flow is also commonly supported by extensive automation in apps, a major plus in managing problem size challenges. Still, even with these advances, closing on some proofs will continue to require hands-on effort and discussion with design and other verification teams.

Getting to formal signoff[75]

Ultimately you and your formal team should assume responsibility for signing-off meaningful components of the verification plan. To do that, you first need to measure progress against a set of goals. We suggest something like the following table as a starting point, though you can certainly adapt (and evolve) this to best suit your needs. These metrics range from simple but relatively low confidence to more complex with increasing degrees of confidence towards signoff quality. Formal tools should help your team gather these statistics. And of course, you will want to trend these statistics during the evolution of projects.

Metric	Degree of confidence provided	Current project status	
		Example target	Actual
Property density	Low	1 assertion / 20 lines of code	
COI/Lines covered by assertions	Medium	100%	
Logic/lines covered by formal core	High	95%	
% (mutated) faults covered by assertions	Highest	90%	

Metrics to use in assessing progress to signoff

In common with simulation, formal tools typically support metrics based on (RTL) line, condition/branch, signal toggle and FSM state coverage. You can select any one of these metrics (in the

---

[75] Here we are not talking about formally signing off a complete block or IP. That is another interesting topic, but beyond the scope of this book.

table we use line-based metrics as an example) to give you a measure of completeness of coverage when performing analysis. The value of these metrics is to provide a much more detailed and functional assessment of assertion coverage than you got in early assertion-density analysis and is a much more concrete measure of verification progress than the simpler assertion-density metrics.

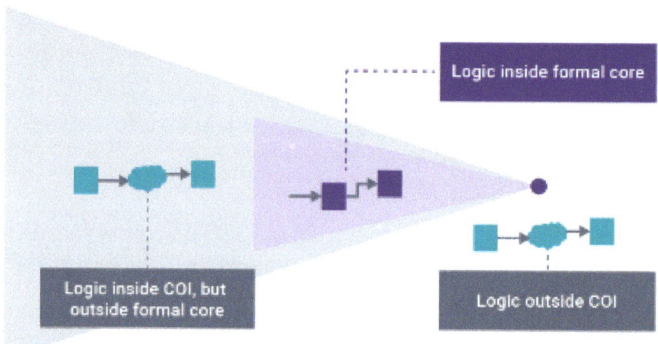

A formal core is usually smaller than the cone of influence

A more precise measure of formal coverage is based on ***formal cores*** (also known as proof cores). A formal core is the portion of logic **essential** to prove a property and is generally a subset of the cone of influence for that property.

All these coverage metrics provide a way to cross-check between what you (or the designer) believe should be touched in an adequate proof versus what you find the tool tells you was really touched. If these correspond, your confidence goes up. If not, perhaps the test needs to be further refined with more assertions or cover points, or perhaps some constraints should be relaxed.

Another very useful cross-check applies RTL mutations to insert bugs into the DUT (this is provided in the Synopsys VC Formal FTA app). You would expect that, under a mutation, at least some assertions should fail. If they don't, this suggests that assertion coverage should be improved, or perhaps that proofs are over-constrained. By running through a number of mutation runs[76] and correcting any exposed coverage problems, signoff quality should become even more secure, delivering the confidence you ultimately want in those proofs.

How you ultimately choose to define signoff based on these metrics should not, conceptually, be very different from how you do this for dynamic signoff. We're just more accustomed to the process for dynamic verification. First you need metrics – we've already discussed these. You want to know that all the assertions you can prove have been proven with no open counter examples, and that cases which resist formal proving are passed back to the dynamic team for a different angle of attack.

You want to know that for each of the proven cases you have good coverage, especially for the strictest levels of checking (formal core) and that no proofs were over-constrained. You should have similar expectations for bounded proofs, with the added requirement that each bound is validated by the designer and/or reachability tests for appropriate cover points. And finally, you will trend metrics to determine where no further progress is being made in formal checking – that you have done *enough* within the bounds of the tasks you assigned to formal.

Again, not so very different from how we approach signoff in dynamic verification. Of course, you will want to build experience and confidence, to decide where you want to set the bar. When you get there, formal signoff should give more confidence than dynamic signoff.

---

[76] Synopsys Certitude is one generally accepted tool for generating mutations.

Chapter 10 Looking Forward

Now you understand the basics of formal verification, where can this technology take you next? Formal continues to be a very active domain in research and experimental applications in industry. Here we mention a range of possibilities discussed often in conferences and papers, ranging from relatively near-term opportunities to longer term but also exciting ideas.

Extending formal methods to C/C++ specifications

In the datapath chapter earlier in this book we talked about reading C/C++ specifications as a reference for equivalence checking RTL implementations. Formal methods and applications first grew around software verification, which became a foundation for equivalence checking against hardware.

Equivalence checking for math hardware implementations works because math functions have software reference models which are widely accepted as golden. Other non-math functions also commonly have software models used for architectural analysis, but we can't apply equivalence checking because those models aren't considered golden, perhaps because they are buggy, more commonly because they are insufficiently precise for equivalence checking.

Why not then run property checking on those software models to formally validate their behavior, which could in turn enable equivalence checking with the corresponding RTL? Conceptually this should not be a big step. We've already noted that formal methods for software are well-established. If formal verification can be moved up to software specifications for hardware for a wider range of functions (not only math functions) that capability could enable a major shift in hardware verification.

Cloud-based formal verification

Cloud-based deployments are becoming increasingly popular in engineering development and validation. Engineering applications tend to stretch every aspect of compute performance, often requiring large numbers of very compute and memory intensive jobs. However, the demand is very spiky. Requirements in early stages of architectural and block development may be quite manageable in an on-premises datacenter, though some complex unit-level formal proofs may still be best handled through a large number of parallel compute engines. Later in SoC design, repeated regressions will periodically explode demand.

This is where the elasticity of cloud deployments, whether hybrid or pure cloud, can offer a huge advantage. Long individual jobs can be compressed by fanning out to many servers, allowing for say a 100 hour run across 10 servers to be flattened to a 10 hour run across 100 servers. With further tool innovation, we might be able to flatten further to a 1 hour run across 1000 servers! Burst demand in regressions can also be handled by elastically expanding resource during the burst, rather than drawing out schedules to fit within a limited capacity on-premises compute farm.

Machine learning and AI

Machine learning and AI have significant impact in many domains today by exploiting learning to accelerate tasks and to improve quality of results. It might surprise you to learn that in hardware verification, these methods first became popular in the formal domain.

Remember the chapter on "What's the catch"? Formal methods must sometimes apply multiple techniques to get to a proof. Apps have greatly simplified this process for the verification engineer, but tools still must work through extended sequence of strategies, which can take a long time. In regressions much of that work will be redundant, repeating similar searches for the best strategy over and over again. Optimizing this process is a natural application for machine learning. Looking back at previous designs runs to learn patterns and find an improved solution.

Another potential application is for root cause detection. For all the work it might take, sometimes finding a bug is the easy part of a verification problem. Figuring out the root cause – the manual debug cycle to find the core design problem that led to the bug – is often more challenging. Automated root cause analysis to help simplify that debug task could also benefit from machine learning.

Architectural formal verification

Some problems cry out for formal verification because the number of cases that must be considered is so high. But the system is far too big to fit in a formal proof and there are no simple candidates to abstract to reduce the proof size to a reasonable level. One good example can be found in cache-coherence checking where proofs span between multiple compute cores and their respective caches.

Multi-core architectures have become popular in SoCs for many reasons. However, to optimize performance, cores depend on local cache memories which provide faster local access to data than would be possible through main memory, while still syncing with main memory as needed.

Those cores still behave as if they are dealing with one logical memory model – the main memory. This gets tricky when two or more cores are working through their respective caches with the same memory address, say address X. If core A updates the value at address X in its cache memory, then core B reads the value at address X in its cache memory, the value core B reads will be wrong. It ought to get the updated value but is unaware of the core A update so instead reads the outdated value in its own cache.

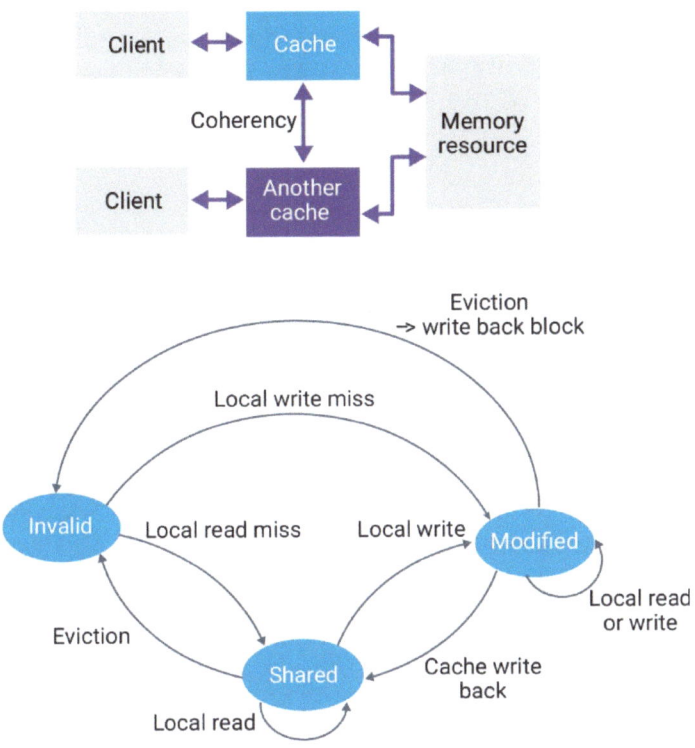

(Simplified) state diagram for cache coherence behavior

Cache-coherent architectures have clever ways to deal with these cases, which essentially come down to snooping on or otherwise being aware of addresses that have been changed by other cores. That triggers, when needed, an update to ensure that all cores continue to see a consistent (coherent) logical view of the memory. As you might imagine, making this work is not simple when still trying to preserve most of the performance advantages of local caches. Proving correctness of operation in all cases, considering all cores and caches, is a complex control problem which is perfect for formal except for the fact that the whole system is too big.

Approaching problems like this requires architectural formal analysis, a methodology and tools to build abstracted models which still retain the important features of the problem, but which are more accessible to formal methods. This is a very hot area of research.

Cross-leveraging simulation and formal investments

Simulation and formal verification have almost identical goals but different ways of approaching those goals. Are there ways to better exploit investments made in one domain into the other?

To some extent this is already happening, in the tools and in everyday application. Verification engineers are now more conscious of both domains, writing assertions which will work equally well in either. Formal engines run under the hood in simulation to find unreachable states and exclude these from coverage statistics. Simulation can be used to provide a starting point for formal proving from a state that would be otherwise impractical to reach using formal methods alone.

Looking further out you could imagine methodologies to leverage simulation testbench collateral into formal testbenches. As the application of formal methods moves closer to end-to-end verification (of blocks) it would be crazy not to exploit all the work that had been put into building and refining legacy dynamic verification testbenches. That intent could, at least in principle, be carried over to formal verification.

The outlook for formal verification

Five years ago, some may still have looked on formal verification as a niche technology for specialized problems, but that view is now rare. Now, large systems and semiconductor companies see formal as a first-class component in any credible verification strategy. Even more than that, as Kiran hints in his foreword, formal methods are now moving up to the point that they can replace simulation in some domains. And they are starting to contribute to system-level domains where formal was previously considered impractical.

These are heady times for formal verification and for formal teams. Through increasing contribution and higher confidence in business-critical requirements, formal technology is becoming increasingly important to product design and development across the digital design spectrum. Formal engineers are also pretty happy. With proven skills in formal verification, they are in high demand and can tackle some of the best intellectual challenges in design today. Their career opportunities are unbounded!

If You Want to Go Deeper

There are many excellent and more detailed sources of information on formal verification. We have selected just a few we think you might want to follow, either based on continuing our theme of a high-level overview without getting too technical or, in the deeper background selection, information on the beginnings of this field, current state and where it is headed.

More detail for beginners

Erik Seligman, Tom Schubert, and M V Achutha Kiran Kumar: Formal Verification: An Essential Toolkit for Modern VLSI Design

Douglas Perry and Harry Foster: Applied Formal Verification: For Digital Circuit Design

Ashish Darbari and Iain Singleton: Industrial Strength Formal Using Abstractions

And, of course, training from the vendors who supply your formal tools

Deeper background

Ed Clarke: Model Checking (MIT Press)

Rolf Dreschler: Formal System Verification: State-of the-Art and Future Trends

Ken McMillan: Symbolic Model Checking

Malay Ganai: SAT-Based Scalable Formal Verification Solutions

Glossary of Formal Terms

Like all of us, formal experts love their jargon and don't always understand that the rest of us may be confused by their specialized language. Fortunately, there aren't too many of these special terms:

Term	Informal definition
App	A pre-packaged application to make user involvement in checking some specific characteristic of a design much simpler than would be required through custom property checking
Assert	A requirement on logical behavior which can be checked in verification
Assume	A SystemVerilog constraint – see Constraint
Assume-guarantee	A way to simplify proof problems is to break the circuit into smaller pieces, say an upstream piece of logic and a downstream piece. You first constrain the inputs to the downstream piece (assume a certain behavior) and prove that piece functions as expected, then you use those assumptions as assertions on the outputs of the upstream piece and prove (guarantee) that those assumptions used in the downstream proof are valid.
BDD	Binary decision diagrams – a data-structure used in certain proof-engines such as for Symbolic Model Checking.
BMC	Bounded model checking – a type of formal engine which checks a property against the circuit in a breadth-first approach until either

	a counter example is found, or a specified depth is reached.
Bounded proof	A case where no proof or counter examples were found out to a specified proof depth, where checking stopped. The proof is bounded because there is no guarantee that counter examples do not exist beyond that depth.
CEX	See counter example
Constraint	A property which limits behavior of some set of signals in the circuit during proving. This could be as simple as fixing a signal value but can be as complex as a checking property. One example would be to define a one-hot constraint on a set of inputs.
Counter example	An example of an execution path (generally presented as waveforms) which demonstrates that the property that you want to prove is clearly false.
Inconclusive	The result of a formal proof is inconclusive, if the proof cannot be completed or counter example cannot be found within the specified time or memory bound by a given formal method, and therefore you can't conclude the correctness or incorrectness of a property. In the case of BMCs, the proof may terminate at a finite depth without finding a failure.
Model-checking	A formal technique which determines if a defined property or specification holds true for a given design or model.

Proof depth	One definition is the cycle bound reached on an inconclusive property result.
SAT	A type of formal engine which looks for a set of variable assignments which will satisfy the disproof of an assertion. This approach can be very fast since it isn't attempting to globally prove the truth of an assertion. It will stop as soon as a counter example is found, if one exists (within the assigned proof depth).
Sequential depth	The number of clock-cycles required from the start of a proof to reach a certain goal – which might be testing an assertion or reaching a coverage property for example.
State space	The graph of all possible states and transitions in a design.
State space diameter (or radius)	The minimum number of cycles required to reach the farthest reachable state from the starting state.
SVA	SystemVerilog Assertions – the standard format used to express properties, assertions, and constraints.
Vacuous proof	When testing a proposition "X implies Y", if X is false then, by a peculiarity of logic, the proposition is true. Of course, this is meaningless, which is why it is called vacuous. In formal verification this can happen with "if X then Y" assertions. If X is never exercised in proving (perhaps because of an over-constraint), the assertion will be reported as vacuously proved.

Witness | In the case of a property that holds true, a witness is one example of a path which demonstrates that the property is true.

Acknowledgements

While four of us are listed as authors, this book would not have been possible without the hard work, dedication, and support of a much larger team.

Firstly, we owe a huge debt of gratitude to our Synopsys marketing leads (Sonia Montgomery and Jin Zhang) who supported us in all aspects of the book. They were the engines that kept us moving forward over many months: keeping us on track, taking care of graphics, searching for background material, and arranging meetings. They simplified the process tremendously, allowing us to simply focus on the content.

We are immensely grateful to Kiran Kumar for his insightful foreword. As a formal verification leader at Intel, the company that got us all into formal verification, he is uniquely qualified to share just how mainstream formal is today and where it is likely to go next.

We want to thank our Synopsys Verification leaders, Ajay Singh, Latha Venkatachari, and Shankar Krishnamoorthy for their constant support and encouragement for the broader formal verification team and for their faith in this team in executing their vision. We also leaned heavily on the knowledge and experience of the formal verification R&D and AE teams. Their insights added practical wisdom and background to this work.

Finally, thanks are due to our families for their infinite patience in supporting us over the many long hours, evenings and weekends we consumed instead of spending time with them.

About the Authors

Bernard Murphy – SemiWiki

Bernard Murphy is a part-time blogger and author with SemiWiki, author of "The Tell-Tale Entrepreneur" and a co-author of the first edition of this book. He is also a content marketing advisor and serves on the board of Mother Lode Wildlife Care in the California Gold Country. Earlier he was CTO at Atrenta and has held technical contributor, management, sales, and marketing roles variously at Cadence, National Semiconductor, Fairchild, and Harris Semiconductor. Bernard has published over 700 articles in industry journals under his own byline. He received his BA in Physics and D. Phil in Nuclear Physics from the University of Oxford.

Manish Pandey – Synopsys

Manish Pandey is vice president R&D and Fellow at Synopsys, and an Adjunct Professor at Carnegie Mellon University and a co-author of the first edition of this book. He completed his PhD in Computer Science from Carnegie Mellon University and a B. Tech. in Computer Science from the Indian Institute of Technology Kharagpur. He currently leads the R&D teams for formal and static technologies, and machine learning at Synopsys. He previously led the development of several static and formal verification technologies at Verplex and Cadence which are in widespread use in the industry. Manish has been the recipient of the IEEE Transaction in CAD Outstanding Young author award and holds over two dozen patents and refereed publications.

Rajeev Ranjan – Synopsys

Dr. Rajeev Ranjan is group director of Applications Engineering at Synopsys for its flagship formal verification product, Synopsys VC Formal. Prior to Synopsys, Dr. Ranjan spent 10+ years at Jasper Design Automation in the role of Chief Technology Officer and VP of Application Engineering developing Jasper, followed by 5+ years at Cadence Design Systems in customer engagement and business development in all areas of verification. Dr. Ranjan is a long-term veteran in formal design verification, has published several papers and holds numerous patents in this area. Dr. Ranjan received his bachelor's degree from IIT Kanpur, master's degree from Urbana Champaign, and PhD from UC Berkeley. He also holds MBA degree from Wharton School of Business, University of Pennsylvania.

Sean Safarpour – Synopsys

Sean Safarpour is a group director of R&D at Synopsys, where he leads the development of VC Formal and all its Apps. Prior to this, he led the Application Engineering team for VC Formal and worked closely with customers to solve some of their most challenging verification problems with the help of formal verification. He is also a co-author of the first edition of this book. Prior to Synopsys, Sean was Director of R&D at Atrenta focused on new technology, and VP of Engineering and CTO at Vennsa Technologies, a start-up focused on automated root-cause analysis using formal techniques. Sean received his PhD from the University of Toronto where he completed his thesis entitled "Formal Methods in Automated Design Debugging".

www.ingramcontent.com/pod-product-compliance
Lightning Source LLC
Chambersburg PA
CBHW040055250526
45473CB00041B/36